Number Theory

Ramanujan Mathematical Society Lectures Notes Series

Vol. 1: Number theory

Vol. 2: The Riemann zeta function and related themes

Vol. 3: Formal language aspects of natural computing

Vol. 4: Commutative algebra and combinatorics

Vol. 5: Convexity in discrete structures

Vol. 6: Number theory and discrete geometry

Vol. 7: Discrete mathematics

Vol. 8: Lectures on operator theory

Vol. 9: Essays on geometric group theory

Vol. 10: Techmüller theory and moduli problems

Vol. 11: Perspectives in geometry and topology

Vol. 12: The Iwasawa theory of totally real fields

Vol. 13: Advances in discrete mathematics and applications

Vol. 14: Ramanujan Rediscovered

Vol. 15: Number Theory

Vol. 16: Symmetry: A Multi-Disciplinary Perspective

Ramanujan Mathematical Society

Lecture Notes Series

Volume 15

Number Theory

Proceedings in Honor of
Prof. T.C. Vasudevan on his Sixtieth
Birthday

Volume editors

M. Manickam
B. Ramakrishnan

International Press
www.intlpress.com

Ramanujan Mathematical Society
Lecture Notes Series, Volume 15
Number Theory

Volume editors:
M. Manickam
B. Ramakrishnan

Published in 2012 by International Press, Somerville, Massachusetts, U.S.A. under license from the Ramanujan Mathematical Society.

This work is also published in the Republic of India, exclusively by the Ramanujan Mathematical Society.

ISBN: 978-1-57146-246-6

Printed in the United States of America.

16 15 14 13 12 1 2 3 4 5 6 7 8 9

Preface

This volume represents the proceedings of the 'International conference on Number Theory and Applications' held during December 2006 organised in the Ramakrishna Mission Vivekananda College, Chennai.

Both of us (editors) were graduate students of Professor T. C. Vasudevan at the Vivekananda College during 1984–89 and so we decided to honour him by bringing out the proceedings of this conference on his sixtieth birthday (during 2007). We invited articles from some of the speakers of the conference and also from among the well-wishers in the mathematical community. The response was very positive and we received twelve articles and our heartfelt thanks to all the authors who contributed to this volume. All the articles were refereed by experts in this field and we also thank the referees for their timely help and support in the process. Due to some unforeseen reasons there was some delay in the publication of this proceedings and our special thanks to the authors for their continued support and patience.

It is our pleasure to thank the RMS Lecture Notes series editorial board for accepting to publish this proceedings under their Lecture Notes series. Finally we thank the Harish-Chandra Research Institute (especially the Special Years in Mathematics project of the XI plan) and the Ramakrishna Mission Vivekananda College for partial financial support to bring out this proceedings.

October 2011

M. Manickam
B. Ramakrishnan

Members of the Advisory Board

Contents

TCV Proceedings No. 15, 2011, pp. 1–5.

Simplices with Monochromatic Vertices: Reflections on Some Problems in Euclidean Ramsey Theory

Sukumar Das Adhikari[1] and Purusottam Rath[2]

[1]Harish-Chandra Research Institute, Chhatnag Road, Jhusi, Allahabad 211 019, India
[2]Chennai Mathematical Institute, Plot H1, SIPCOT IT Park, Padur PO, Siruseri 603 103, India
e-mail: adhikari@mri.ernet.in; purusottam.rath@gmail.com

Abstract. We here discuss some questions in Euclidean Ramsey Theory. More precisely, these are related to a conjecture suggested by Gurevich which says that for any finite colouring of the Euclidean plane $\mathbb{R}^2$, there always exists a triangle of unit area with monochromatic vertices.

Keywords. Gurevich's conjecture, Monochromatic configurations.

2000 Subject Classification: 05D10

Dedicated to Prof. T. C. Vasudevan on the Occasion of his 60th Birthday

1. Introduction

The general philosophy underlying the mathematics of Ramsey Theory is that any substructure of a structured object retains certain regularity properties of the original structure once the size of the substructure is 'large'. In the present exposition, we discuss a question belonging to the realm of *Euclidean Ramsey Theory*. Here, given a group H of transformations of the standard Euclidean space $\mathbb{R}^n$, a finite subset K of $\mathbb{R}^n$ and a positive integer r; one considers the question of whether for any partition of $\mathbb{R}^n$ into r disjoint parts, there exists an image of K under an element of H which falls into one of these disjoint r classes. Using the language of colouring which is commonplace in this set up, for any r colouring of the n-dimensional Euclidean plane, one investigates the (possible) existence of a monochromatic image of K under the action of some element of H.

The subject of Euclidean Ramsey Theory was developed in the pioneering works of Erdős, Graham, Montgomery, Rothschild, Spencer and Strauss ([7], [8], [9]). Just to give a sample of the theorems in this area, we have a theorem due to Frankl and Rödl [10] which asserts the following: given a triangle ABC and an integer r, $r \geq 2$, for n sufficiently large and an arbitrary r-coloring of $\mathbb{R}^n$, one can find a monochromatic copy of ABC.

We proceed to state the fundamental results obtained by Erdős, Graham, Montgomery, Rothschild, Spencer and Strauss ([7], [8], [9]). We need the following definitions.

Definitions. *Given a finite set $C \subset \mathbb{R}^n$, we shall say that $\mathbf{R}(C, n, r)$ holds, if for any r-colouring of $\mathbb{R}^n$, there exists a monochromatic set $C' \subset \mathbb{R}^n$ such that C' is congruent to C. Here, by saying that C' is congruent to C, one means that C' is the image of C under some element in the group of Euclidean motions of $\mathbb{R}^n$.*

Given a finite subset C in some Euclidean space, one says that C is *Ramsey* if for every positive integer r, there is a positive integer $n = n(C, r)$ such that $\mathbf{R}(C, n, r)$ holds. With this terminology, the result of Frankl and Rödl mentioned above can be expressed by saying that 'all triangles are Ramsey'.

It is known that if $C_1 \subset \mathbb{R}^{n_1}$ and $C_2 \subset \mathbb{R}^{n_2}$ are two finite sets which are Ramsey, then so is $C_1 \times C_2$. Since any two-point set is Ramsey, it follows that the Cartesian product of finitely many two-point sets, referred to as *bricks* in the literature, is also Ramsey. Another fundamental result is that if a set S is not spherical (that is, S is not contained in the surface of a sphere), then it can not be Ramsey.

It is an important open question to determine whether every spherical set is Ramsey. For further details and other important results in this area, we refer to [1] and [13].

The ethos of our present note originates from a conjecture suggested by Gurevich which says that for any finite colouring of the Euclidean plane $\mathbb{R}^2$, there always exists a triangle of unit area with monochromatic vertices. Note that there is no restriction on the shape of the triangle. Graham [12] not only settled this particular question, but proved the following stronger version where he managed to specify even the shape of the triangles of prescribed area having monochromatic vertices.

Theorem 1 (Graham). *For a positive integer r, there is a positive integer $T(r)$ such that for any given r-colouring of the set of integer lattice points in $\mathbb{R}^2$, there is always a right angled triangle of area $T(r)$ such that all its vertices have the same colour.*

Later, exploiting the main idea of Graham, a shorter proof of the above theorem was furnished by the first author [2]. From the proof itself, one observes that the triangle in the above theorem can be made to satisfy further constraints of having two of its sides parallel to the axes and by suitable scaling, the following corollary is easily obtained.

Corollary. *For any $\alpha > 0$ and any pair of non-parallel lines L_1 and L_2, in any partition of the plane into finitely many classes, some class contains the vertices of a triangle which has area α and two sides parallel to the lines L_i.*

As has been mentioned in [13], the natural generalisation of the above result for higher dimensional simplices can be obtained by similar arguments.

Apart from mentioning several interesting known results related to Theorem 1, the paper [12] of Graham raised the question of establishing results similar to his theorem for other configurations which are not simplexes, such as vertices of a parallelogram (or of a rhombus or of a rectangle) in the plane.

In a recent work [4], the present authors considered the corresponding result in the case of vertices of a trapezium. More precisely, the following theorem has been proved.

Theorem 2. *Let $\mathbb{R}^2$ denote the Euclidean plane endowed with rectangular Cartesian co-ordinates. For a positive integer r, there is a positive integer T_r such that given any*

$\delta > 0$ and an r-colouring of the set of integer lattice points in $\mathbb{R}^2$, there is always a trapezium of area $d\delta$ with $d \leq T_r$ such that all of its four vertices are lattice points of the same colour and three of its sides are parallel to the co-ordinate axes.

In the present note, we explain the underlying principles involved in the proof of the above theorems with brief descriptions of the required pre-requisites. There are lot of important configurations like parallelograms and rectangles which appear to elude the above techniques. However, it is possible that the above techniques may be refined to address these configurations and this present exposition is motivated to highlight these techniques. We also suggest some other problems related to the question of Gurevich. There are very few results in this set up with a lot of unanswered questions.

2. Sketch of proofs of the theorems

In what follows, $\mathbb{Z}^+$ will denote the set of positive integers and $\mathbb{Z}_{\geq 0}$ the set of non-negative integers. The heart of the proof rests on the following theorem of van der Waerden. This result has been instrumental in a series of developments in Combinatorics and Number Theory. We refer to a recent article [3] by the first author for an expository account of this beautiful theorem.

Theorem (van der Waerden). *Given any $k, r \in \mathbb{Z}^+$, there exists a number $W = W(k,r) \in \mathbb{Z}^+$ such that for any r-colouring of the set $\{1, 2, \ldots, W(k,r)\}$, there is a monochromatic arithmetic progression (A.P) of k terms.*

We now sketch the proof of Theorem 1 following [2]. The proof is by induction on r. When $r = 1$, simply take any right angled triangle of desired shape with area 1. For $r \geq 2$, assume that there are positive integers C_{r-1}, $D_{r-1} (\geq 2)$ and T_{r-1} such that for any $(r-1)$-colouring of the lattice points in the rectangle $\{(a,b) : 0 \leq a, b \leq C_{r-1} - 1\}$, there is a monochromatic triangle of desired shape with area T_{r-1} having vertices in this set.

We define the following numbers:

$$E_{r-1} := D_{r-1}!T_{r-1}, \quad w_r := W(E_{r-1}(C_{r-1}+1), r).$$

Now for any r-colouring of $\mathbb{R}^2$, we consider the set of points $(nT_{r-1}, 0)$ with $n = 1, 2, \ldots$ and using van der Waerden's theorem, retrieve a monochromatic set S, of colour say r, lying on the X-axis such that

$$S = \{(a_r + ig_rT_{r-1}, 0) : 1 \leq i \leq E_{r-1}(C_{r-1}+1)\} \subset \{(nT_{r-1}, 0) : n \leq w_r\}.$$

The idea now is the following, we consider a suitable region above the X-axis such that the projection of any point in this region on the X-axis lies in our retrieved set S. If any point in this region has colour r, we consider the projection lying in S and then choose another suitable point in S to construct a right angled triangle with colour r. But if no point in this region has colour r, then the region has received an $(r-1)$-colouring and the idea is to consider a scaled sub-region and apply the induction hypothesis so that the area obtained in both the cases match. We refer to [2] for the exact details.

We note that in the theorem of Graham, we were interested in constructing monochromatic right angled triangles of desired area and whence we required two degrees of freedom, one in the vertical direction and one in the horizontal direction. The theorem of van der Waerden ensured these two degrees of freedom. Now for constructing trapeziums of desired area, clearly we require three degrees of freedom, either two along vertical and one along horizontal or vice versa. For this, a natural idea would be to exploit the following two dimensional case of a theorem due to G. Grünwald (Gallai) which is a higher dimensional generalization of van der Waerden's theorem. We refer to [3] or [5] for a proof.

Theorem (Grünwald). *Let $r \in \mathbb{Z}^+$. Then, given any finite set $S \subset (\mathbb{Z}_{\geq 0})^2$, and an r-colouring of $(\mathbb{Z}_{\geq 0})^2$, there exist positive integers $N = N(S, r)$ and 'd' and a point 'v' in $(\mathbb{Z}_{\geq 0})^2$ such that the set $dS + v$ is monochromatic where $dS + v$ is a subset of $\{B_N \stackrel{\text{def}}{=} \{(a, b) : a, b \in \mathbb{Z}_{\geq 0}, 0 \leq a, b \leq N\}$.*

3. Concluding remarks

(1) As mentioned before, it is not known whether Theorem 2 holds for important configurations like parallelograms and rectangles. It is easy to see that it can not hold for squares. Note that in these cases, we will need four degrees of freedom, two each in both horizontal as well as vertical direction. Perhaps one can consider some other generalization of van der Waerden's theorem. Other possibility is to apply such a theorem in the three-dimensional space and consider appropriate projections in the plane.
(2) The same questions can also be raised about other constrained configurations rather than those with a given area. It has been noted by E. G. Straus that Theorem 1 can not hold for triangles with a prescribed perimeter. For if it did, then the corresponding corollary would hold for perimeter 1. But colouring the plane in a multicoloured cyclic check board pattern with squares of side $1/16$, we see that it is not true.
(3) One can also consider higher dimensional analogs of the questions suggested by Graham. To begin with, it may be worthwhile to investigate what results can be derived using the theorems of van der Waerden as well as that of Grünwald.
(4) We also mention the question asked by Graham related to the possible growth of the function $T(r)$; from the proofs of Theorem 1 in [12] and [2], one may expect that it will be related to that of the function $W(k, r)$. It should be mentioned that while the recent years have seen significant improvement of the upper bound for $W(k, r)$ as one of the consequences of a very deep work [11] of Gowers, there remains a big gap between this and the available lower bounds (see [15] or [16], for instance). We also refer to a recent article [6] of Dumitrescu and Jiang which, among other things, gives a very short proof of the original Gurevich conjecture (with no restriction on the shape of the triangle). Finally, we recommend the enjoyable article of Graham [14] for some other important questions and results in Euclidean Ramsey Theory.

References

[1] S. D. Adhikari, Aspects of Combinatorics and Combinatorial Number Theory, Narosa Publishing House (2002).

[2] S. D. Adhikari, On a theorem of Graham, *Expo. Math.*, **19(4)** (2001) 365–367.

[3] S. D. Adhikari, A pearl of Number Theory, the theorem of van der Waerden, *Mathematics Newsletter, Published by Ramanujan Mathematical Society*, **18(2)** (2008) 58–62.

[4] S. D. Adhikari, Purusottam Rath, Monochromatic configurations for finite colourings of the plane, *Note di Matematica*, **22(1)** (2003/04) 59–73.

[5] P. G. Anderson, A generalisation of Baudet's conjecture (van der Waerden's Theorem), *Amer. Math. Monthly*, **83(5)** (1976) 359–361.

[6] A. Dumitrescu and M. Jiang, Monochromatic simplices of any volume, *Discrete Math.*, **310(4)** (2010) 956–960. (An earlier version appeared in the Proceedings of the 20th Canadian Conference on Computational Geometry, (CCCG 2008), 71–74).

[7] P. Erdős, R. L. Graham, P. Montgomery, B. L. Rothschild, J. H. Spencer and E. G. Straus, Euclidean Ramsey Theorems I, *J. Combinatorial Theory Ser. A*, **14** (1973) 341–363.

[8] P. Erdős, R. L. Graham, P. Montgomery, B. L. Rothschild, J. H. Spencer and E. G. Straus, Euclidean Ramsey Theorems II, Infinite and finite sets, 520–557. *Colloq. Math. Soc. Janos Bolyai*, **10**, Keszthely, Hungary and North Holland, Amsterdam (1975).

[9] P. Erdős, R. L. Graham, P. Montgomery, B. L. Rothschild, J. H. Spencer and E. G. Straus, Euclidean Ramsey Theorems III, Infinite and finite sets, 559–583. *Colloq. Math. Soc. Janos Bolyai*, **10**, Keszthely, Hungary and North Holland, Amsterdam (1975).

[10] P. Frankl and V. Rödl, All triangles are Ramsey, *Trans. Amer. Math. Soc.*, **297(2)** (1986) 777–779.

[11] W. T. Gowers, A new proof of Szemerédi's theorem, *Geom. Funct. Anal.*, **11(3)** (2001) 465–588.

[12] R. L. Graham, On partitions of E^n, *J. Combinatorial Theory Ser. A*, **28** (1980) 89–97.

[13] R. L. Graham, Rudiments of Ramsey Theory, American Mathematical Society, Providence, R.I., (1981).

[14] R. L. Graham, Recent trends in Euclidean Ramsey theory, *Discrete Math.*, **136(1–3)** (1994) 119–127.

[15] R. L. Graham, On the growth of a van der Waerden-like function, *Integers*, **6** (2006) A 29.

[16] P. R. Herwig, M. J. H. Heule, P. M. van Lambalgen and H. van Maaren, A new method to construct lower bounds for van der Waerden numbers, *Electron. J. Combin.*, **14(1)** (2007) R 6.

TCV Proceedings No. 15, 2011, pp. 7–16.

On the Fourier-Jacobi-Coefficients of Eisenstein Series of Klingen Type

Siegfried Böcherer

Institut für Mathematatik, Universität Mannheim, 68131 Mannheim, Germany
e-mail: boecherer@math.uni-mannheim.de

Abstract. The Fourier-Jacobi coefficients of a Klingen Eisenstein series can be decomposed into an Eisenstein series part and a cuspidal part. We describe some properties of this decomposition in the simplest case of a Klingen Eisenstein series of degree two.

Keywords. Jacobi forms, Fourier-Jacobi-expansion, Klingen-Eisenstein series.

0. Introduction

The well known decomposition of the space of (elliptic) modular forms into Eisenstein series and cusp forms was generalized to Siegel modular forms by Maaß [12] and Klingen [9], using Poincare series and Klingen-Eisenstein series (respectively) (at least for large weights and for the full modular group). We may write the space M_n^k of Siegel modular forms of degree n and weight k as a direct sum

$$M_n^k = \bigoplus_{r=0}^{n} M_{n,r}^k,$$

where $M_{n,r}^k$ corresponds to the space of cusp forms of degree r (via an iterated ϕ-operator).

The same type of decomposition was given by Dulinski [5] in the context of Jacobi forms; it is more complicated than in the Siegel case, if the index is not a squarefree number.

A Siegel modular form has a Fourier-Jacobi-expansion and it is natural to ask how the decompositions above fit together. It is trivial that Fourier-Jacobi-coefficients of Siegel cusp forms are cuspidal Jacobi forms. We also know from [2] that the Fourier-Jacobi-coefficients of Siegel-Eisenstein series are (linear combinations of) Jacobi-Eisenstein series. The aim of this note is to investigate the question above for the first very simple not yet understood case, namely for Klingen Eisenstein series $E_{2,1}^k(f)$ of degree two attached to elliptic cusp forms f (i.e. for the space $M_{2,1}^k$). We do not aim at any general statement: We consider only the level one case, we stick to squarefree indices and we focus on Fourier coefficients associated to maximal lattices (which is almost the same as saying that the discriminant of the corresponding positive binary quadratic form is a fundamental discriminant). We hope to come back to these questions in greater generality on a future occasion. At this point however, it is not clear to this author whether one can obtain general results along the lines of this paper (general means: For Siegel modular forms of arbitrary degree and level and Fourier-Jacobi-expansions for indices of arbitrary size).

It turns out that a Fourier-Jacobi-coefficient of a Klingen-Eisenstein series $E^k_{2,1}(f)$ has a nonzero Jacobi-Eisenstein part and a nonzero cuspidal Jacobi part. We identify their Fourier coefficients in terms of subseries of the formula for the Fourier coefficients of $E^k_{2,1}(f)$.

A sometimes important feature in the theory of modular forms is that Fourier coefficients of Eisenstein series grow (in general) much faster than the Fourier coefficients of cusp forms. Indeed, for the Klingen Eisenstein series $E^k_{2,1}(f)$ the decomposition of its Fourier coefficients in two parts by the observation above is a decomposition into two parts with different growth (at least along certain Fourier coefficients). This is relevant for questions about representation numbers of quadratic forms and has some overlap with old investigations of Raghavan and myself [3].

1. On Klingen-Eisenstein series

For the basic setup of Siegel modular forms we refer the reader to [10] or [7]. We let $Sp(n,\mathbb{R})$ act in the usual way on Siegel's upper half space $\mathbb{H}_n$. We also use the action of $Sp(n,\mathbb{R})$ on functions $F : \mathbb{H}_n \longrightarrow \mathbb{C}$ defined by $(F|_k g)(Z) := \det(CZ+D)^{-k}F(g\langle Z\rangle)$ for $g = \left(\begin{smallmatrix} A & B \\ C & D\end{smallmatrix}\right)$. We define the space M^k_n (S^k_n respectively) of degree n Siegel modular forms (cusp forms respectively) of weight k in the usual way. We denote by $C_{n,r}$ the parabolic subgroup of $Sp(n,\mathbb{R})$ defined by the vanishing of the entries in the last $n-r$ rows and first $n+r$ colums. Starting from a cusp form $f \in S^k_r$ we define the Klingen Eisenstein series $E^k_{n,r}(f)$ to be the element of M^k_n defined by

$$E^k_{n,r}(f,Z) := \sum_{M\in C_{n,r}(\mathbb{Z})\backslash Sp(n,\mathbb{Z})} f^*|_k M,$$

where f^* is the function defined on $\mathbb{H}_n$ by $f^*(Z) := f(Z_1)$ with Z_1 denoting the $r\times r$ block in the upper left corner of Z. Then we put

$$M^k_{n,r} := \{E^k_{n,r}(f) | f \in S^k_r\}.$$

The Fourier expansion of Klingen-Eisenstein series is well-known [1,2,13]. We only need here the case of

$$E^k_{2,1}(f) = \sum_T a(f,T)e^{2\pi i\ tr(T\cdot Z)},$$

where T runs over half-integral positive-semidefinite symmetric matrices of size 2, $tr(..)$ denotes the trace of a matrix and $f \in S^k_1$ is a normalized Hecke eigenform defined by the Fourier series

$$f(\tau) = \sum_{n=1}^{\infty} b(f,n)e^{2\pi i n\tau}.$$

The formulas simplify considerably if we look at the *primitive* Fourier coefficients $a(f,T)^*$ defined (for $T>0$) by the relation (see [3])

$$a(f,T)=\sum_W a(f,T[W^{-1}])^*,$$

where W runs over $GL(2,\mathbb{Z})\backslash GL(2,\mathbb{Q})\cap\mathbb{Z}^{2,2}$. Note that this is actually a finite sum. From the papers quoted above, we have then

$$a\left(f,\begin{pmatrix} t & 0\\ 0 & 0\end{pmatrix}\right)=b(f,t),$$

and for $T>0$

$$a(f,T)^*=c_{2,1}\frac{a_2^k(T)^*}{L_2(f,2k-2)}\sum\frac{1}{2}\frac{b(f,n)A(T,n)}{n^{k-1}}.$$

Here $A(T,n)$ denotes the number of integral representations of n by the quadratic form T,

$$c_{2,1}=\frac{\zeta(1-k)\zeta(2k-2)}{2},$$

and $a_2^k(T)^*$ denotes the primitive Fourier coefficient of the Siegel-Eisenstein series $E_{2,0}^k$, which is defined by

$$E_{2,0}^k(Z)=\sum_{\gamma\in C_{2,0}(\mathbb{Z}\backslash Sp(2,\mathbb{Z})}1|_k\gamma=\sum_{C,D}\det(CZ+D)^{-k}=\sum_T a_2^k(T)e^{2\pi i tr(TZ)},$$

where (C,D) runs over all non-associate primitive symmetric pairs. Finally,

$$L_2(f,s):=\prod_p\frac{1}{(1-\alpha_p^2p^{-s})(1-\beta_p^2p^{-s})(1-p^{k-1-s})}$$

is the symmetric square L-function attached to the normalized eigenform f with $\alpha_p+\beta_p=b(f,p)$, $\alpha_p\beta_p=p^{k-1}$.

2. The Jacobi-Eisenstein part and the Jacobi-cuspidal part of $E_{2,1}^k(f)$

After decomposing $Z\in\mathbb{H}_2$ in the form

$$Z=\begin{pmatrix}\tau & z\\ z & \tau'\end{pmatrix}\quad(\tau,\tau'\in\mathbb{H},z\in\mathbb{C})$$

we can read off the Fourier-Jacobi expansion

$$E_{2,1}^k(f)=\sum_m\phi_m(\tau,z)e^{2\pi i m\tau'}$$

from the formulas of the previous section by collecting the appropriate Fourier coefficients. We tacitly assume throughout that $m>0$.

Then ϕ_m can be decomposed into a Jacobi-Eisenstein series and a cusp form:

$$\phi_m = \mathcal{E}_{k,m} + \phi_m^o.$$

Attached to this decomposition, there is a natural decomposition of the corresponding Fourier coefficients, which we want to study for $T = \begin{pmatrix} n & \frac{r}{2} \\ \frac{r}{2} & m \end{pmatrix}$:

$$a(f, T) = e_m(n, r) + c_m(n, r),$$

where $e_m(n, r)$ and $c_m(n, r)$ are the Fourier coefficients of the Eisenstein part and the cuspidal part of ϕ_m (respectively):

$$\mathcal{E}_{k,m}(\tau, z) = \sum_{n,r} e_m(n, r) e^{2\pi i(n\tau + rz)}$$

$$\phi_m^o(\tau, z) = \sum_{n,r} c_m(n, r) e^{2\pi i(n\tau + rz)}.$$

We consider now the special case

$$m \text{ squarefree.}$$

This condition implies that there is only one cusp (in the sense of Jacobi forms). For such m

$$\mathcal{E}_{k,m} = a\left(f, \begin{pmatrix} 0 & 0 \\ 0 & m \end{pmatrix}\right) \cdot E_{k,m} = b(f, m) \cdot E_{k,m},$$

where

$$E_{k,m}(\tau, z) = \sum_{\gamma \in G_\infty^J(\mathbb{Z}) \backslash G^J(\mathbb{Z})} 1|_{k,m}\gamma = \sum_{n,r} a_{k,m}(n, r) e^{2\pi i(n\tau + rz)}$$

is the (unique normalized) Jacobi-Eisenstein series in the sense of [6], G^J denotes the Jacobi group and $|_{k,m}$ is the slash-operator from the theory of Jacobi forms.

We recall from [2] that the mth Fourier-Jacobi coefficient of $E_{2,0}^k$ is equal to

$$c_{1,0}\sigma_{k-1}(m) \cdot E_{k,m},$$

where

$$c_{1,0} = \frac{2}{\zeta(1-k)}.$$

This implies for the Fourier coefficients

$$c_{1,0} \cdot \sigma_{k-1}(m) \cdot a_{k,m}(n, r) = a_2^k\left(\begin{pmatrix} n & \frac{r}{2} \\ \frac{r}{2} & m \end{pmatrix}\right).$$

For the Fourier expansion of $\mathcal{E}_{k,m}$ and ϕ_m^o this means (at least if T is a maximal lattice)

$$e_m(n, r) = b(f, m) \frac{a_2^k(T)}{c_{1,0}\,\sigma_{k-1}(m)}$$

$$c_m(n, r) = a(f, T) - b(f, m) \frac{a_2^k(T)}{c_{1,0}\,\sigma_{k-1}(m)}.$$

Our aim is to show that $e_m(n,r)$ and $c_m(n,r)$ can be identified with nice subseries of the series defining $a(f,T)$:

We study the subseries of

$$\sum_{n=1}^{\infty} \frac{b(f,n)A(T,n)}{n^{k-1}} = \sum_{u,v\in\mathbb{Z}}{}' \frac{b(f,nu^2+ruv+mv^2)}{(mu^2+ruv+nv^2)^{k-1}}$$

defined by the condition $u=0$: We get

$$2\sum_{v=1}^{\infty} \frac{b(f,mv^2)}{(mv^2)^{k-1}} = 2\frac{b(f,m)}{\sigma_{k-1}(m)} \sum_{u=1}^{\infty} \frac{b(f,u^2)}{u^{2k-2}}$$

$$= 2\frac{b(f,m)}{\sigma_{k-1}(m)} \frac{L_2(f,2k-2)}{\zeta(2k-2)}.$$

Here we have used the multiplicative properties of the Fourier coefficients of the eigenform f and the wellknown relation (see e.g. [17]) between the Rankin-L-function $\sum_{m=1}^{\infty} b(f,m^2)m^{-s}$ and the symmetric-square L-function $L_2(f,s)$.

From the simple identity

$$c_{2,1}\frac{a_2^k(T)}{L_2(f,2k-2)} \times \frac{b(f,m)}{\sigma_{k-1}(m)} \frac{L_2(f,2k-2)}{\zeta(2k-2)} = b(f,m)\frac{a_2^k(T)}{c_{1,0}\,\sigma_{k-1}(m)}$$

we see that

we have now identified the Jacobi-Eisenstein part of the Fourier Jacobi-coefficient of $E_{2,1}(f)$ to be given by the subseries defined by $u=0$.

We summarize this as main result of this paper:

Proposition. *Let $f \in S_1^k$ be a normalized Hecke eigenform. The Fourier-Jacobi coefficients ϕ_m of $E_{2,1}^k(f)$ decompose as $\phi_m = \mathcal{E}_{k,m} + \phi_m^0$. If the discriminant of $T = \begin{pmatrix} n & \frac{r}{2} \\ \frac{r}{2} & m \end{pmatrix}$ is fundamental and m is squarefree, then the (n,r)-Fourier coefficients in this decomposition are given by*

$$e_m(n,r) = c_{2,1}\frac{a_2^k(T)}{L_2(f,2k-2)}\frac{1}{2}\sum_{v\in\mathbb{Z}\setminus\{0\}} \frac{b(f,mv^2)}{(mv^2)^{k-1}} = a_2^k(T)\cdot\frac{b(f,m)}{c_{1,0}\sigma_{k-1}(m)}$$

$$c_m(n,r) = c_{2,1}\frac{a_2^k(T)}{L_2(f,2k-2)}\frac{1}{2}\sum_{0\neq u,v\in\mathbb{Z}} \frac{b(f,nu^2+ruv+mv^2)}{(nu^2+ruv+mv^2)^{k-1}}.$$

We remark that the formula above for $e_m(n,r)$ is linear (valid not only for eigenforms!).

3. On growth properties of ϕ_m^o and $\mathcal{E}_{k,m}$

We call a symmetric positive definite matrix $T = \begin{pmatrix} n & \frac{r}{2} \\ \frac{r}{2} & m \end{pmatrix}$ reduced, if it satisfies $|r| \leq m \leq n$. For such reduced matrices we tacitly use in the sequel the well known properties

$$m = \text{Min}\,\{nu^2 + ruv + mv^2 | (u, v) \in \mathbb{Z}^2 \setminus \{(0,0)\}\}$$
$$\leq \sqrt{\frac{4}{3}} \cdot \det(T)^{\frac{1}{2}}$$

and

$$\frac{1}{2}\begin{pmatrix} n & 0 \\ 0 & m \end{pmatrix} \leq \begin{pmatrix} n & \frac{r}{2} \\ \frac{r}{2} & m \end{pmatrix} \leq \begin{pmatrix} n & 0 \\ 0 & m \end{pmatrix}.$$

Here $S \leq T$ means for two such matrices that $T - S$ is positive semidefinit.

Concerning the growth of the Fourier coefficients we remind the reader of the well-known fundamental fact (see [8]) that

$$a_n^k(T) \sim \det(T)^{k - \frac{n+1}{2}} \qquad (k > n + 1)$$

and from this we have then (always under the condition that m is squarefree and $\det(2T)$ is fundamental)

$$e_m(n, r) \sim \frac{|b(f, m)|}{m^{k-1}} \det(T)^{k - \frac{3}{2}}.$$

The (best possible) upper bound for $|b(f, m)|$ is Deligne's estimate (Ramanujan-Petersson conjecture); for results about $\limsup |b(f, m) m^{-\frac{k-1}{2}}|$ we refer to [14,16].

It is now an interesting point to give an estimate for $c_m(n, r)$. We mention that the estimate

$$c_m(n, r) \ll_{k,\epsilon} \left(1 + \frac{\det(T)^{\frac{1}{2}+\epsilon}}{m}\right)^{\frac{1}{2}} \frac{\det(T)^{\frac{k}{2}-\frac{3}{4}}}{m^{\frac{k}{2}-1}} \cdot \|\phi_m^o\|,$$

which is actually valid for any Jacobi cusp form ([11,4], is not sufficient for our purposes. To get a satisfactory result from this, we would need in addition an estimate for the Petersson norms

$$\|\phi_m^o\| = \mathbf{O}(m^\delta) \tag{3.1}$$

of the cuspidal part of the Fourier-Jacobi expansion of $E_{2,1}^k(f)$ with an appropriate δ. Indeed, in the case of reduced T we could get from the estimates above

$$|c_m(n, r)| = \mathbf{O}(\det(T)^{\frac{k}{4}+\frac{\delta}{2}-\frac{1}{4}+\epsilon}).$$

Therefore, any such estimate with $\delta < \frac{3k-5}{2}$ would give an estimate $|c_m(n, r)| = \mathbf{O}(\det(T)^\kappa)$ with $\kappa < k - \frac{3}{2}$. Such a uniform estimate for $\|\phi_m^o\|$ seems not to be available at the moment.

Remark. We remind the reader that in the case of Siegel cusp forms such a uniform estimate is known, e.g. a Hecke type bound: $\|\phi_m\| = \mathbf{O}(m^{\frac{k}{2}})$, which leads to an estimate of Fourier coefficients of Siegel cusp forms of degree 2 by $\mathbf{O}(\det(T)^{\frac{k}{2}-\frac{1}{4}})$. For a slightly better estimate, valid for Siegel cusp forms of arbitrary degree, see [4].

Due to the lack of an estimate (3.1), we apply a different strategy: Using an estimate for $b(f, m)$ in the formula for $c_m(n, r)$, we consider more generally a sum of type (for an arbitrary real number α with $\alpha > 2$)

$$\sum_{0 \neq u, v \in \mathbb{Z}} (nu^2 + ruv + mv^2)^{-\alpha}.$$

Note that this sum is not a class invariant of the quadratic form $T = \begin{pmatrix} n & \frac{r}{2} \\ \frac{r}{2} & m \end{pmatrix}$.

Lemma. *Assume that T is reduced ($|r| \leq m \leq n$). Then*

$$\sum_{0 \neq u, v \in \mathbb{Z}} (nu^2 + 2ruv + mv^2)^{-\alpha} = \mathbf{O}(\det(T)^{-\frac{\alpha}{2}}) \tag{3.2}$$

For our purposes any negative exponent of $\det(T)$ *would be sufficient.*

Proof. It is sufficient to consider the case of a diagonal form. We start from the elementary estimate

$$\sum_{u=1}^{\infty} (u^2 + C)^{-\alpha} \leq c_\alpha \cdot C^{-\alpha + \frac{1}{2}},$$

valid for any $C > 0$ with $c_\alpha = \int_0^\infty (x^2 + 1)^{-\alpha} dx$. This follows from comparing the sum with an integral of type $\int_0^\infty (x^2 + C)^{-\alpha} dx$. We omit the details.

Then we first consider for $A, B > 0$

$$\sum_{u, v \in \mathbb{Z} \setminus \{0\}} (Au^2 + Bv^2)^{-\alpha} = A^{-\alpha} \sum_{u, v \in \mathbb{Z} \setminus \{0\}} \left(u^2 + \frac{B}{A} v^2\right)^{-\alpha}$$

$$\leq 4 c_\alpha A^{-\frac{1}{2}} B^{-\alpha + \frac{1}{2}} \zeta(2\alpha - 1).$$

To get an estimate symmetric in A and B we interchange the roles of A and B and take the geometric mean of the two estimates to obtain

$$\sum_{u, v \in \mathbb{Z} \setminus \{0\}} (Au^2 + Bv^2)^{-\alpha} = \mathbf{O}(AB)^{-\frac{\alpha}{2}}.$$

It remains to estimate the partial series defined by $v = 0$:

$$\sum_{u=1}^{\infty} (nu^2)^{-\alpha} = n^{-\alpha} \cdot \zeta(2\alpha) = \mathbf{O}(\det(T)^{-\frac{\alpha}{2}}).$$

Here we used that $n \geq \det(T)^{\frac{1}{2}}$ for diagonal reduced T.

Remark. With a little more effort, one gets for not necessarily reduced T a more general version of the estimate (3.2)

$$= \mathbf{O}(\det(T)^{-\alpha + \frac{1}{2}} \cdot \min\{n, m\}^{\alpha - 1} + n^{-\alpha}).$$

Keeping in mind that a reduced matrix is of same size as its diagonal (in the sense of quadratic forms) we obtain (using Deligne's estimate for $b(m, f)$)

Proposition. *Assume that T is reduced (and $-\det(2T)$ fundamental and m square-free), then*

$$|c_m(n,r)| = \mathbf{O}(\det(T)^{\frac{3k}{4}-\frac{5}{4}+\epsilon}).$$

This means that indeed (at least for these special T) we see that the growth of the Fourier coefficients is of different order for $c_m(n,r)$ and $e_m(n,r)$. Note that if we use Hecke's estimate instead of Deligne's we get $\frac{3k}{4}-\frac{3}{2}$ in the exponent, which is enough for our purpose.

4. Application to representation numbers

We start from an even unimodular lattice L of rank $2k$ in the Euclidean space $\mathbb{R}^{2k}$. The associated theta series of degree n is then

$$\theta^n(L,Z) := \sum_{X\in\mathbb{Z}^{(2k,n)}} e^{2\pi i tr(S[X]Z)}.$$

Here we have tacitly identified the lattice L with a positive definite even unimodular symmetric matrix $2S$. This theta series defines an element of M_n^k with Fourier expansion

$$\theta^n(L,Z) = \sum_T A(S,T)e^{2\pi i tr(TZ)},$$

where the Fourier coefficients are representation numbers

$$A(S,T) = \sharp\{X\in\mathbb{Z}^{2k,n}|S[X]=T\}.$$

The direct sum $M_2^k = M_{2,0}\oplus M_{2,1}^k\oplus M_{2,2}^k$ gives rise to a decomposition

$$\theta_2(L) = \theta_{2,0}(L)+\theta_{2,1}(L)+\theta_{2,2}(L) \quad (\theta_{2,i}(L)\in M_{2,i}^k).$$

This also implies a corresponding decomposition of the representation number $A(S,T)$ into Fourier coefficients of the $\theta_{2,i}(L)$:

$$A(S,T) = A_{2,0}(S;T)+A_{2,1}(S,T)+A_{2,2,}(S,T).$$

Moreover, we can decompose $A_{2,1}(S,T)$ still further according to the decomposion $\phi_m = \mathcal{E}_{k,m}+\phi_m^o$:

$$A_{2,1}(S,T) = A_{2,1}^{\mathcal{E}}(S,T)+A_{2,1}^{o}(S,T).$$

What kind of information about the asymptotic behaviour of $A(S,T)$ can we get from these decompositions? Note that the case of fixed minimum of T will be of particular interest.

To study the $M_{2,1}^k$-part of $\theta_2(L)$ in detail, we need the degree one decomposition

$$\theta_1(L) = E_{1,0}^k+\theta_{1,1}(L) \quad (\theta_{1,1}(L)\in S_1^k),$$

in particular,

$$b(\theta_{1,1}(L), m) = A(S, m) - c_{1,0}\,\sigma_{k-1}(m).$$

Then the formula for $e_m(n, r)$ given in section 2 implies

$$\begin{aligned} A_{2,1}(S, T)^{\mathcal{E}} &= \frac{1}{c_{1,0}\,\sigma_{k-1}(m)}(A(S, m) - c_{1,0}\,\sigma_{k-1}(m))a_2^k(T) \\ &= \frac{a_2^k(T)}{c_{1,0}\,\sigma_{k-1}(m)} A(S, m) - a_2^k(T). \end{aligned}$$

Now we recall that $A_{2,0}(S, T) = a_2^k(T)$ and we see that the contribution $A_{2,2}(S, T)$ gets cancelled against a part of $A^{\mathcal{E}}_{2,1}(S, T)$:

$$\begin{aligned} A(S, T) &= A_{2,0}(S, T) + A^{\mathcal{E}}_{2,1}(S, T) + A^{o}_{2,1}(S, T) + A_{2,2,}(S, T) \\ &= 0 + \frac{a_2^k(T)}{c_{1,0}\,\sigma_{k-1}(m)} \cdot A(S, m) + A^{0}_{2,1}(S, T) + A_{2,2}(S, T). \end{aligned}$$

We recall that these formulas are true if $-\det(2T)$ is a fundamental discriminant and m is squarefree; the relevant estimates are

$$a_2^k(T) \sim \det(T)^{k-\frac{3}{2}}$$

$$A^{0}_{2,1}(S, T) = \mathbf{O}(\det(T)^{\frac{3k}{4}-\frac{5}{4}})$$

and

$$A_{2,2}(S, T) = \mathbf{O}(\det(T)^{\frac{k}{2}-\frac{1}{4}+\epsilon}).$$

Note that the second estimate was given under the additional assumption that T is reduced.

It is remarkable that we have now a "dominating" term which is zero iff $A(S, m) = 0$; in this respect this formula is not so much different from Theorem IV in [3] (proved there in much greater generality). However, our estimates are all in terms of powers of $\det(T)$, whereas in [3] we emphasized the case $T = \begin{pmatrix} n & \frac{r}{2} \\ \frac{r}{2} & m \end{pmatrix}$ with r, m fixed and n growing.

Note (added in proof). In his ongoing PhD-thesis Thorsten Paul (Universität des Saarlandes) will investigate this topic further.

Acknowledgements

I thank R. Schulze-Pillot for many discussions about the subject of this paper.

References

[1] S. Böcherer, Über gewisse Siegelsche Modulformen zweiten Grades, *Math. Ann.*, **261** (1981) 23–41.

[2] S. Böcherer, Über die Fourier-Jacobi-Entwicklung Siegelscher Eisensteinreihen, *Math. Z.*, **183** (1984) 21–46.

[3] Böcherer, Raghavan, On Fourier Coefficients of Siegel modular forms, *J. Reine angew. Math.*, **384** (1988) 80–101.
[4] S. Böcherer and W. Kohnen, Estimates for Fourier coefficients of Siegel cusp forms, *J. Reine angew. Math.*, **297** (1983) 499–517.
[5] J. Dulinski, A decomposition theorem for Jacobi forms, *Math. Ann.*, **303** (1995) 473–498.
[6] M. Eichler and D. B. Zagier, The Theory of Jacobi Forms, Birkhäuser (1985).
[7] E. Freitag, Siegelsche Modulfunktionen, Springer.
[8] Y. Kitaoka, Lectures on Siegel modular forms and representations of quadrtaic forms, *Tata Inst. of Fund. Research*, Bombay (1986).
[9] H. Klingen, Zum Darstellungssatz für Siegelsche Modulformen, *Math. Z.*, **102** (1967) 30–43.
[10] H. Klingen, Introductory Lectures on Siegel Modular Forms, Cambridge University Press.
[11] W. Kohnen, Estimates for Fourier coefficients of Siegel cusp forms of degree 2, *Compos. Math.*, **87** (1992) 231–240.
[12] H. Maaß, Über die Darstellung der Modulformen n-ten Grades durch Poincaresche Reihen, *Math. Ann.*, **123** 125–151.
[13] S. Mizumoto, Fourier coefficients of generalized Eisenstein series of degree 2, *I. Invent. Math.*, **65** (1981) 115–135.
[14] R. A. Rankin, Fourier coefficients of cusp forms, *Math. Proc. Camb. Phil. Soc.*, **100** (1986) 5–29.
[15] R. Schulze-Pillot, Local conditions for global representations of quadratic forms, *Acta Arith.*, **138** (2009) 289–299.
[16] J.-P. Serre, Repartitions des valeurs propres de T_p, *J. Amer. Math. Soc.*, **10** (1997) 75–102.
[17] D. Zagier, Modular forms whose Fourier coefficients involve zeta-functions attached to quadratic fields, In: Modular Functions in One Variable VI. Lecture Notes in Math. 627, Springer (1977).

TCV Proceedings No. 15, 2011, pp. 17–30.

The Theta-Operator and the Divisors of Modular Forms

Sanoli Gun[1] and B. Ramakrishnan[2]

[1]The Institute of Mathematical Sciences, CIT Campus, Taramani, Chennai 600 113, India
[2]Harish-Chandra Research Institute, Chhatnag Road, Jhunsi, Allahabad 211 019, India
e-mail: sanoli@imsc.res.in; ramki@hri.res.in

Abstract. In 2004, Bruinier, Kohnen and Ono proved a relation between the values of a sequence of modular functions at points in the divisor of a meromorphic modular form on $SL_2(\mathbb{Z})$. In this article, we prove similar results in the case of modular functions on certain genus zero congruence subgroups.

Keywords. Modular functions, divisor, theta operator, Genus zero congruence subgroups.

2000 Subject Classification: Primary: 11F03; Secondary: 11F11, 11F13

In Honour of Professor Vasudevan's 60th Birthday

1. Introduction

The values of modular functions have significant impact on the theory of modular forms. For instance, the values of j function at quadratic irrationals play an important role in the work of R. Borcherds [3] on infinite product expansions of certain modular functions. In 2002, D. Zagier [14] showed a relation between traces of such j values with Fourier coefficients of a meromorphic modular form of half-integral weight. In this direction, J. H. Bruinier, W. Kohnen and K. Ono [4] proved a relation between the values of a certain sequence of modular functions (which are derived from the classical j function) and arithmetic of the coefficients of modular forms on $SL_2(\mathbb{Z})$. They also gave an explicit description of the action of Ramanujan's theta-operator on modular forms, universal recurrences for the coefficients of certain meromorphic modular forms and an identity for the exponents in Borcherds infinite product expansions.

Ono [11, Problem 4.32, p. 83] proposed the problem of proving similar results for other types of modular forms. Several authors [1,2,5,6,8] have worked on this problem and they have extended the above results for some genus zero congruence subgroups of $SL_2(\mathbb{Z})$ and Hecke groups. In this paper, we consider the rest of the genus zero congruence subgroups of $SL_2(\mathbb{Z})$ and prove analogous results.

2. Preliminaries and statement of results

Let $\mathcal{H}$ be the complex upper half-plane and $q = e^{2\pi i z}$, $z \in \mathcal{H}$. Let $E_2(z)$ be the Eisenstein series of weight 2 on $SL_2(\mathbb{Z})$ given by $E_2(z) = 1 - 24\sum_{n\geq 1}\sigma(n)q^n$, where $\sigma(n)$ denotes the sum of the divisors of n. Ramanujan's theta-operator is defined by

$$\theta f(z) = \frac{1}{2\pi i}\frac{d}{dz} f(z) = \sum_{n=h}^{\infty} n a(n) q^n,$$

where $f(z) = \sum_{n=h}^{\infty} a(n)q^n$ is a meromorphic modular form. Moreover, we say that a meromorphic modular form f is normalized if $a(h) = 1$.

It is known [12] that for $N = 1, \ldots, 10, 12, 13, 16, 18, 25$, the congruence subgroup $\Gamma_0(N)$ has genus zero. As mentioned before, the cases corresponding to $N = 1$ was considered in [4], $N = 2, 3, 5, 7, 13$ was considered in [1] and the case $N = 4$ was considered in [2]. So, we consider the remaining cases. Let $\mathcal{G} := \{6, 8, 9, 10, 12, 16, 18, 25\}$. For $N \in \mathcal{G}$, define modular functions $\phi_N(z)$ on $\Gamma_0(N)$ as follows:

$$\phi_6(z) = \frac{\eta^5(z)\eta(3z)}{\eta(2z)\eta^5(6z)}, \qquad \phi_8(z) = \frac{\eta^4(z)\eta^2(4z)}{\eta^2(2z)\eta^4(8z)},$$

$$\phi_9(z) = \frac{\eta^3(z)}{\eta^3(9z)}, \qquad \phi_{10}(z) = \frac{\eta^3(z)\eta(5z)}{\eta(2z)\eta^3(10z)},$$

$$\phi_{12}(z) = \frac{\eta^3(z)\eta(4z)\eta^2(6z)}{\eta^2(2z)\eta(3z)\eta^3(12z)}, \qquad \phi_{16}(z) = \frac{\eta^2(z)\eta(8z)}{\eta(2z)\eta^2(16z)},$$

$$\phi_{18}(z) = \frac{\eta^2(z)\eta(6z)\eta(9z)}{\eta(2z)\eta(3z)\eta^2(18z)}, \qquad \phi_{25}(z) = \frac{\eta(z)}{\eta(25z)}.$$

It is easy to see that these functions are weight zero univalent modular functions having a simple pole at infinity, simple zero at zero and holomorphic at all other cusps. See [7]. Further, for each such N, we define a sequence of meromorphic modular forms $\{j_{n,N}(z)\}_{n=1}^{\infty}$ as follows: $j_{n,N}(z)$ is the unique meromorphic modular form on $\Gamma_0(N)$ which is holomorphic on $\mathcal{H}$, vanishes at the cusp 0 and has the following Fourier expansion at infinity

$$j_{n,N}(z) = q^{-n} + c_{n,N}(0) + c_{n,N}(1)q + c_{n,N}(2)q^2 + \cdots . \qquad (1)$$

Such modular functions exist because they can be constructed as polynomials in $\phi_N(z)$ with no constant term. For example,

$$j_{1,6}(z) = \phi_6(z) = q^{-1} - 5 + 6q + 4q^2 - 3q^3 - 12q^4 - \cdots ,$$

$$j_{2,6}(z) = \phi_6(z)^2 + 10\phi_6(z) = q^{-2} - 13 + 8q + 30q^2 + 24q^3 - 36q^4 - \cdots ,$$

$$j_{2,6}(z) = \phi_6(z)^3 + 15\phi_6(z)^2 + 57\phi_6(z) = q^{-2} - 23 - 9q + 36q^2 + 186q^3 + 144q^4 - \cdots ,$$

$$\vdots$$

Table 1.

N	$(\alpha_{1,N}, \alpha_{2,N}, \ldots, \alpha_{\sigma_0(N),N})$	N	$(\alpha_{1,N}, \alpha_{2,N}, \ldots, \alpha_{\sigma_0(N),N})$
6	$\left(\frac{-5}{24}, \frac{1}{12}, \frac{-1}{8}, \frac{5}{4}\right)$	12	$\left(\frac{-1}{8}, \frac{1}{6}, \frac{1}{8}, \frac{-1}{6}, \frac{-1}{2}, \frac{3}{2}\right)$
8	$\left(\frac{-1}{6}, \frac{1}{6}, \frac{-1}{3}, \frac{4}{3}\right)$	16	$\left(\frac{-1}{12}, \frac{1}{12}, 0, \frac{-1}{3}, \frac{4}{3}\right)$
9	$\left(\frac{-1}{8}, 0, \frac{9}{8}\right)$	18	$\left(\frac{-1}{12}, \frac{1}{12}, \frac{1}{8}, \frac{-1}{4}, \frac{-3}{8}, \frac{3}{2}\right)$
10	$\left(\frac{-1}{8}, \frac{1}{12}, \frac{-5}{24}, \frac{5}{4}\right)$	25	$\left(\frac{-1}{24}, 0, \frac{25}{24}\right)$

Uniqueness of these functions follows by observing that the difference of two such functions satisfying (1) would be a holomorphic modular function, vanishing at the cusp 0 (and holomorphic at other cusps), and therefore equal to the zero function. Note that for $n = 1$, we have $j_{1,N}(z) = \phi_N(z)$. Also, for $\tau \in \mathcal{H}$, we define

$$H_{\tau,N}(z) = \sum_{n=1}^{\infty} j_{n,N}(\tau)q^n. \tag{2}$$

Throughout the paper $t_1 = 1, t_2, \ldots, t_{\sigma_0(N)} = N$ denote the divisors of N written in the increasing order, where $\sigma_0(N)$ denotes the number of divisors of N. For $N \in \mathcal{G}$, let $-s_i/t_i$, where $s_i < t_i$, $\gcd(s_i, t_i) = 1$, $t_i|N$, $t_i \neq 1, N$ be the inequivalent cusps of $\Gamma_0(N)$, which are not equal to the cusps ∞ and 0. Note that the cusp 0 corresponds to the divisor $t_i = 1$ and the cusp ∞ to the divisor $t_i = N$. With these notations, we now state the main results of this paper. The first theorem gives an explicit formula for the function $H_{\tau,N}(z)$.

Theorem 1. *Let $N \in \mathcal{G}$. Then for $\tau \in \mathcal{H}$, we have*

$$H_{\tau,N}(z) = \frac{\phi_N(\tau)}{\phi_N(z) - \phi_N(\tau)} \sum_{j=1}^{\sigma_0(N)} \alpha_{j,N} E_2(t_j z),$$

where the values of $\alpha_{j,N}$ are given in Table 1. In particular, $H_{\tau,N}(z)$ is a meromorphic modular form of weight 2 on $\Gamma_0(N)$.

For $N \in \mathcal{G}$, let $\mathcal{F}_N$ denote a fundamental domain for the action of $\Gamma_0(N)$ on $\mathcal{H}$ (which does not include the cusps), which is the usual non-compact Riemann surface $Y_0(N) = \Gamma_0(N)\backslash\mathcal{H}$. If $\tau \in \mathcal{H}$, then define $l_{\tau,N} \in \{1, 2\}$ as the order of the isotropy subgroup of τ in $\Gamma_0(N)/\{\pm I\}$. If $\tau \in \mathcal{H}$ and f is a meromorphic modular form on $\Gamma_0(N)$, then define $v_{\tau,N}(f) := \frac{1}{l_{\tau,N}}\mathrm{ord}_\tau(f)$, where $\mathrm{ord}_\tau(f)$ denotes the order of vanishing of f at the point $\tau \in \mathcal{H}$. It is a fact that $v_{\tau,N}(f)$ measures the order of vanishing with respect to the local coordinates on $Y_0(N)$. For details we refer to [13].

The next theorem gives a formula for the coefficients in the infinite product expansion of any meromorphic modular form on $\Gamma_0(N)$, where $N \in \mathcal{G}$.

Table 2. For a fixed value of N, the numbers $h_{t_i,N}$ denote the value of f at the cusp $-s_i/t_i$, $t_i \neq 1, N$. In the table we have dropped the subscript N and have written only h_{t_i}.

N	$(\beta_{1,N}, \beta_{2,N}, \ldots, \beta_{\sigma_0(N),N})$
6	$\left(\frac{5h}{24} + \frac{3h_2}{8} + \frac{h_3}{3} - \frac{k}{4}, -\frac{h}{12} - \frac{3h_2}{4} - \frac{h_3}{3} + \frac{k}{4}, \frac{h}{8} - \frac{3h_2}{8} - h_3 + \frac{k}{4}, -\frac{5h}{4} + \frac{3h_2}{4} + h_3 - \frac{k}{4}\right)$
8	$\left(-\frac{h}{6} + \frac{h_2}{3} + \frac{h_4}{6} - \frac{k}{6}, \frac{h}{6} - h_2 + \frac{h_4}{6} + \frac{k}{6}, -\frac{h}{3} + \frac{2h_2}{3} - \frac{5h_4}{3}, \frac{4h}{3} + \frac{4h_4}{3}\right)$
9	$\left(\frac{h}{8} + \frac{3h_3}{8} - \frac{k}{8}, -\frac{3h_3}{2} + \frac{k}{8}, -\frac{9h}{8} + \frac{9h_3}{8}\right)$
10	$\left(\frac{h}{8} + \frac{5h_2}{24} + \frac{h_5}{6} - \frac{5k}{24}, -\frac{h}{12} - \frac{5h_2}{12} - \frac{h_5}{6} + \frac{5k}{24}, \frac{5h}{24} - \frac{5h_2}{24} - \frac{5h_5}{6} + \frac{5k}{24}, -\frac{5h}{4} + \frac{5h_2}{12} + \frac{5h_5}{12} - \frac{5k}{24}\right)$
12	$\left(\frac{h}{8} + \frac{h_2}{4} + \frac{h_3}{6} + \frac{h_4}{8} + \frac{h_6}{12} - \frac{k}{4}, -\frac{h}{6} - \frac{h_2}{12} - \frac{h_3}{6} + \frac{h_6}{12} + \frac{k}{4}, -\frac{h}{8} - \frac{h_2}{4} - \frac{h_3}{2} - \frac{h_4}{8} + \frac{h_6}{4} + \frac{k}{4}, \frac{h}{6} + \frac{h_2}{2} - \frac{h_4}{2} - \frac{h_6}{6}, \frac{h}{2} + \frac{3h_2}{4} + \frac{h_3}{2} - \frac{7h_6}{4} - \frac{k}{4}, -\frac{3h}{2} - \frac{h_2}{2} + \frac{h_4}{2} + \frac{3h_6}{2}\right)$
16	$\left(\frac{h}{12} + \frac{h_2}{6} + \frac{h_4}{6} + \frac{h_8}{12} - \frac{k}{6}, -\frac{h}{12} - \frac{h_2}{2} + \frac{h_4}{6} - \frac{h_8}{12} + \frac{k}{6}, \frac{h_2}{3} - \frac{5h_4}{3} + \frac{h_8}{3}, \frac{h}{3} + \frac{4h_4}{3} - \frac{5h_8}{3}, -\frac{4h}{3} + \frac{4h_8}{3}\right)$
18	$\left(\frac{h}{12} + \frac{h_2}{8} + \frac{h_3}{4} + \frac{h_6}{8} + \frac{h_9}{12} - \frac{k}{4}, -\frac{h}{12} - \frac{h_2}{4} - \frac{h_3}{4} - \frac{h_9}{12} + \frac{k}{4}, -\frac{h}{8} - \frac{h_2}{8} - h_3 + \frac{h_6}{4} + \frac{k}{4}, \frac{h}{4} + \frac{h_2}{4} + h_3 - \frac{6h_6}{4} - \frac{k}{4}, \frac{3h}{8} + \frac{3h_3}{4} - \frac{3h_6}{8} - \frac{3h_9}{4}, -\frac{3h}{2} - \frac{3h_3}{4} + \frac{3h_6}{2} + \frac{3h_9}{4}\right)$
25	$\left(\frac{h}{24} + \frac{5h_5}{24} - \frac{5k}{48}, -\frac{5h_5}{4} + \frac{5k}{48}, -\frac{25h}{24} + \frac{25h_5}{24}\right)$

Theorem 2. *Let $f(z) = \sum_{n=h}^{\infty} a(n)q^n$ be a normalized meromorphic modular form of weight k on $\Gamma_0(N)$, $N \in \mathcal{G}$. Suppose that $\{c(n)\}_{n=1}^{\infty}$ are complex numbers such that*

$$f(z) = q^h \prod_{n=1}^{\infty} (1 - q^n)^{c(n)}. \tag{3}$$

Then for $n \geq 1$, we have

$$c(n) = \frac{1}{n} \sum_{d \mid n} \mu\left(\frac{n}{d}\right) \sum_{\tau \in \mathcal{F}_N} v_{\tau,N}(f) j_{d,N}(\tau) - 24 \sum_{j=1}^{\sigma_0(N)} \frac{\beta_{j,N}}{t_j} \chi_{\mathbb{Z}}\left(\frac{n}{t_j}\right), \tag{4}$$

where $\beta_{j,N}$ are given by Table 2, and $\chi_{\mathbb{Z}}$ is the characteristic function of the integers.

Remark 1. Note that Lemma 5 ensures the existence of complex numbers $\{c(n)\}$ satisfying (3) for any meromorphic modular form $f(z)$.

Next we give an explicit formula for the action of the theta-operator on meromorphic modular forms on $\Gamma_0(N)$.

Theorem 3. *Suppose that $N \in \mathcal{G}$ and that $f(z) = \sum_{n=h}^{\infty} a(n)q^n$ is a normalized meromorphic modular form of weight k on $\Gamma_0(N)$. Define the weight 2 modular* *function*

$$f_{\theta,N} = \sum_{\tau\in\mathcal{F}_N} \upsilon_{\tau,N}(f)H_{\tau,N}.$$

For $1 \le j \le \sigma_0(N)$, *let* $\beta_{j,N}$ *be constants such that*

$$\frac{\theta f}{f}(z) + \sum_{j=1}^{\sigma_0(N)} \beta_{j,N}E_2(t_j z)$$

is a modular function of weight 2 *on* $\Gamma_0(N)$ *which vanishes at the cusp* ∞ *and* $-s_i/t_i$, $t_i|N$, $t_i \ne 1, N$. *(See Lemma 6 for the existence of such constants.) Then we have*

$$\frac{\theta f}{f}(z) = -f_{\theta,N}(z) - \sum_{j=1}^{\sigma_0(N)} \beta_{j,N}E_2(t_j z).$$

The values of $\beta_{j,N}$, $1 \le j \le \sigma_0(N)$ *are given in Table 2.*

As an application, we prove a theorem which gives a recurrence for the Fourier coefficients of normalized meromorphic modular forms on $\Gamma_0(N)$. For $N \in \mathcal{G}$, $n \ge 1$ and $\beta_{j,N}$, $1 \le j \le \sigma_0(N)$ as in Table 2, define the polynomial

$$\begin{aligned} &F_n(\beta_{1,N}, \ldots, \beta_{\sigma_0(N),N}, x_1, x_2, \ldots, x_{n-1}) \\ &= \sum (-1)^{m_1+\cdots+m_{n-1}} \frac{(m_1+\cdots+m_{n-1}-1)!}{m_1!\ldots m_{n-1}!} x_1^{m_1}\ldots x_{n-1}^{m_{n-1}} \\ &+ \frac{24}{n}\sum_{j=1}^{\sigma_0(N)} \beta_{j,N}\sigma\left(\frac{n}{t_j}\right), \end{aligned}$$

where the first sum is over all non negative integers $m_1, \ldots, m_{n-1}$ with $\sum_{j=1}^{n-1} j \cdot m_j = n$. Here we use the fact that $\sigma(n) = 0$ if $n \notin \mathbb{N}$. Then, we have

Theorem 4. *Suppose that* $N \in \mathcal{G}$ *and that*

$$f(z) = q^h + \sum_{n=h+1}^{\infty} a(n)q^n$$

is a meromorphic modular form of weight k *on* $\Gamma_0(N)$. *For* $n \ge 1$, *let* F_n *be the polynomial defined as above. Then, we have*

$$\begin{aligned} a(h+n) &= F_n(\beta_{1,N}, \ldots, \beta_{\sigma_0(N),N}, a(h+1), a(h+2), \ldots, a(h+n-1)) \\ &\quad - \frac{1}{n}\sum_{\tau\in\mathcal{F}_N} \upsilon_{\tau,N}(f)j_{n,N}(\tau), \end{aligned}$$

where the constants $\beta_{j,N}$ *are as in Table 2. Moreover, if we assume that* f *is holomorphic and non-vanishing on* $\mathcal{H}$, *then we get the following simpler expression:*

$$a(h+n) = \frac{24}{n}\sum_{l=1}^{n} a(h+n-l)\sum_{j=1}^{\sigma_0(N)} \beta_{j,N}\sigma\left(\frac{l}{t_j}\right). \qquad (5)$$

3. Proofs

We first prove Theorem 2 and use it to prove the other theorems. Before that we need some lemmas. To begin with we state a lemma which was proved by Eholzer and Skoruppa [9] (see also [4]).

Lemma 5. *Let* $f(z) = \sum_{n=h}^{\infty} a(n)q^n$ *be a meromorphic modular form in a neighborhood of* $q = 0$ *with* $a(h) = 1$. *Then there are uniquely determined complex numbers* $c(n)$ *such that*

$$f(z) = q^h \prod_{n=1}^{\infty} (1 - q^n)^{c(n)},$$

where the product converges in some neighborhood of $q = 0$. *Moreover, we have the identity*

$$\frac{\theta f}{f} = h - \sum_{n=1}^{\infty} \sum_{d|n} c(d) d q^n.$$

In the next lemma we construct a modular form by applying the theta-operator on the given meromorphic modular form and adding a suitable linear combinations of the Eisenstein series of weight 2. This lemma plays a crucial role in proving Theorem 2.

Lemma 6. *Let* $N \in \mathcal{G}$ *and*

$$f(z) = q^h \prod_{n \geq 1} (1 - q^n)^{c(n)}$$

be a normalized meromorphic modular form of weight k for $\Gamma_0(N)$. *Also let h be the order of vanishing of f at infinity and* $h_{t_i,N}$ *be the order of vanishing at the cusp* $-s_i/t_i$, *where* $s_i < t_i$ *and* $\gcd(s_i, t_i) = 1$, $t_i | N$, $t_i \neq 1, N$. *For* $1 \leq j \leq \sigma_0(N)$, *let* $\gamma_{j,N}$ *satisfy the following equation.*

$$\sum_{j=1}^{\sigma_0(N)} \frac{N}{t_j} \gamma_{j,N} + \frac{kN}{12} = 0. \tag{6}$$

Then the function

$$F(z) = \frac{\theta f}{f}(z) + \sum_{j=1}^{\sigma_0(N)} \gamma_{j,N} E_2(t_j z)$$

is a meromorphic modular form of weight 2 on $\Gamma_0(N)$. *Further, F is holomorphic at the cusps with values* $h + \sum_{i=1}^{\sigma_0(N)} \gamma_{i,N}$ *at* ∞ *and* $\frac{h_{t_i,N}}{w_i} + L_{t_i,N}$ *at the cusp* $-s_i/t_i$, $t_i \neq 1, N$ *respectively, where* w_i *is the width of the cusp* $-s_i/t_i$. *The values of* (s_i, t_i), w_i, *and* $L_{t_i,N}$ *for* $N \in \mathcal{G}$ *are given in Table 3.*

Proof. Suppose that f is a meromorphic modular form of weight k on $\Gamma_0(N)$ and $\gamma \in \Gamma_0(N)$ with (c, d) as the last row. Using the transformation formulas for $f(z)$

Table 3.

N	$(s_i, t_i), t_i \neq 1, N$	w_i	$L_{t_i,N}$
6	$(1,2)$	3	$\gamma_{1,6} + \gamma_{2,6} + \frac{\gamma_{3,6}}{9} + \frac{\gamma_{4,6}}{9}$
6	$(1,3)$	2	$\gamma_{1,6} + \frac{\gamma_{2,6}}{4} + \gamma_{3,6} + \frac{\gamma_{4,6}}{4}$
8	$(1,2)$	2	$\gamma_{1,8} + \gamma_{2,8} + \frac{\gamma_{3,8}}{4} + \frac{\gamma_{4,8}}{16}$
8	$(1,4), (3,4)$	1	$\gamma_{1,8} + \gamma_{2,8} + \gamma_{3,8} + \frac{\gamma_{4,8}}{4}$
9	$(1,3), (2,3)$	1	$\gamma_{1,9} + \gamma_{2,9} + \frac{\gamma_{3,9}}{9}$
10	$(1,2)$	5	$\gamma_{1,10} + \gamma_{2,10} + \frac{\gamma_{3,10}}{25} + \frac{\gamma_{4,10}}{25}$
10	$(1,5)$	2	$\gamma_{1,10} + \frac{\gamma_{2,10}}{4} + \gamma_{3,10} + \frac{\gamma_{4,10}}{4}$
12	$(1,2)$	3	$\gamma_{1,12} + \gamma_{2,12} + \frac{\gamma_{3,12}}{9} + \frac{\gamma_{4,12}}{4} + \frac{\gamma_{5,12}}{9} + \frac{\gamma_{6,12}}{36}$
12	$(1,3)$	4	$\gamma_{1,12} + \frac{\gamma_{2,12}}{4} + \gamma_{3,12} + \frac{\gamma_{4,12}}{16} + \frac{\gamma_{5,12}}{4} + \frac{\gamma_{6,12}}{16}$
12	$(1,4)$	3	$\gamma_{1,12} + \gamma_{2,12} + \frac{\gamma_{3,12}}{9} + \gamma_{4,12} + \frac{\gamma_{5,12}}{9} + \frac{\gamma_{6,12}}{9}$
12	$(1,6)$	1	$\gamma_{1,12} + \gamma_{2,12} + \gamma_{3,12} + \frac{\gamma_{4,12}}{4} + \gamma_{5,12} + \frac{\gamma_{6,12}}{4}$
16	$(1,2)$	4	$\gamma_{1,16} + \gamma_{2,16} + \frac{\gamma_{3,16}}{4} + \frac{\gamma_{4,16}}{16} + \frac{\gamma_{5,16}}{64}$
16	$(1,4), (3,4)$	1	$\gamma_{1,16} + \gamma_{2,16} + \gamma_{3,16} + \frac{\gamma_{4,16}}{4} + \frac{\gamma_{5,16}}{16}$
16	$(3,8)$	1	$\gamma_{1,16} + \gamma_{2,16} + \gamma_{3,16} + \gamma_{4,16} + \frac{\gamma_{5,16}}{4}$
18	$(1,2)$	9	$\gamma_{1,18} + \gamma_{2,18} + \frac{\gamma_{3,18}}{9} + \frac{\gamma_{4,18}}{9} + \frac{\gamma_{5,18}}{81} + \frac{\gamma_{6,18}}{81}$
18	$(1,3), (2,3)$	2	$\gamma_{1,18} + \frac{\gamma_{2,18}}{4} + \gamma_{3,18} + \frac{\gamma_{4,18}}{4} + \frac{\gamma_{5,18}}{9} + \frac{\gamma_{6,18}}{36}$
18	$(1,6), (5,6)$	1	$\gamma_{1,18} + \gamma_{2,18} + \gamma_{3,18} + \gamma_{4,18} + \frac{\gamma_{5,18}}{9} + \frac{\gamma_{6,18}}{81}$
18	$(1,9)$	2	$\gamma_{1,18} + \frac{\gamma_{2,18}}{4} + \gamma_{3,18} + \frac{\gamma_{4,18}}{4} + \gamma_{5,18} + \frac{\gamma_{6,18}}{4}$
25	$(1,5), (2,5), (3,5), (4,5)$	1	$\gamma_{1,25} + \gamma_{2,25} + \frac{\gamma_{3,25}}{25}$

and $E_2(z)$, we get

$$E_2(t_i \gamma z) = (cz+d)^2 E_2(t_i z) - \frac{6ic}{t_i \pi}(cz+d), \; t_i | N,$$

$$\frac{\theta f}{f}(\gamma z) = (cz+d)^2 \frac{\theta f}{f}(z) - \frac{kci}{2\pi}(cz+d).$$

Now using the condition (6) it is easy to see that

$$F(\gamma z) = (cz+d)^2 F(z).$$

This shows that F is a meromorphic modular form of weight 2 on $\Gamma_0(N)$. We illustrate the calculation of the value at the cusps with one example. The other cases follow in a similar way. Let γ be an element of the full modular group $SL_2(\mathbb{Z})$ which sends the cusp ∞ to the cusp $-s_i/t_i$, $t_i|N$, $t_i \neq 1, N$. Put $f|_k\gamma(z) = g(z)$. Then it is easy to see that

$$\frac{\theta f}{f}\bigg|_2 \gamma = \frac{\theta g}{g} - \frac{kic}{2\pi}(cz+d)^{-1},$$

where (c, d) is the last row of the matrix γ. We also have the following transformation property of $E_2(z)$:

$$E_2|_2\gamma(z) = E_2(z) - \frac{6ic}{\pi}(cz+d)^{-1}.$$

Let us consider the case $N = 6$ and the cusp $-1/2$, where $t_i = 2$. Let $\gamma = \left(\begin{smallmatrix} -1 & -a \\ 2 & b \end{smallmatrix}\right)$, where $2a - b = 1$ be the corresponding matrix. Then it is easy to get the following transformations:

$$E_2(z)|_2\gamma(z) = E_2(z) - \frac{12i}{\pi}(2z+b)^{-1}$$

$$E_2(2z)|_2\gamma(z) = E_2(2z) - \frac{6i}{\pi}(2z+b)^{-1}$$

$$E_2(3z)|_2\gamma(z) = \frac{1}{9}E_2(z/3) - \frac{4i}{\pi}(2z+b)^{-1}$$

$$E_2(6z)|_2\gamma(z) = \frac{1}{9}E_2(2z/3) - \frac{2i}{\pi}(2z+b)^{-1}.$$

The width of the cusp $-1/2$ is 3 and the order of vanishing of f at the cusp $-1/2$ is $h_{2,6}$. So, the constant term in the expansion of $\frac{\theta g}{g}$ is $h_{2,6}/3$. By the condition (6) we see the remainder terms in the above transformations cancel and the value of F at the cusp $-1/2$ is equal to the constant term of $\frac{\theta g}{g}$ plus $L_{2,6}$, the constant term of the linear combination $\gamma_{1,6}E_2(z) + \gamma_{2,6}E_2(2z) + \gamma_{3,6}E_2(3z) + \gamma_{4,6}E_2(6z)$. In other words, the value of F at the cusp $-1/2$ is

$$\frac{h_{2,6}}{3} + \gamma_{1,6} + \gamma_{2,6} + \frac{\gamma_{3,6}}{9} + \frac{\gamma_{4,6}}{9}.$$

(See the first row of Table 3.) □

Lemma 7. *Let $N \in \mathcal{G}$ and*

$$f(z) = q^h \prod_{n\geq 1}(1-q^n)^{c(n)}$$

be a normalized meromorphic modular form of weight k for $\Gamma_0(N)$. Also let $h_{t_i,N}$ be the order of vanishing of f at the cusp $-s_i/t_i$, $t_i \neq 1, N$. Let $\beta_{j,N}$ for $1 \leq j \leq \sigma_0(N)$ be the constants as in Theorem 3 (in other words, the constants $\beta_{j,N}$ be such that the function $\frac{\theta f}{f} + \sum_{j=1}^{\sigma_0(N)} \beta_{j,N}E_2(t_jz)$ is a modular form of weight 2 on $\Gamma_0(N)$ which vanishes at all the cusps except at 0). Then for $m \geq 1$, we have

$$\sum_{\tau\in\mathcal{F}_N} v_{\tau,N}(f)j_{m,N}(\tau) = \sum_{d|m} c(d)d + 24\sum_{j=1}^{\sigma_0(N)} \beta_{j,N}\sigma\left(\frac{m}{t_j}\right).$$

Proof. The proof follows in a similar fashion as done in the earlier works ([1,2,4]). However, for completeness we give a brief sketch of the proof.

For $N \in \mathcal{G}$, let $X_0(N)$ be the compact Riemann surface obtained by adjoining the cusps of $\Gamma_0(N)$ to $Y_0(N) = \Gamma_0(N)\backslash\mathcal{H} = \mathcal{F}_N$. If G is a meromorphic modular form of weight 2 on $\Gamma_0(N)$, then on the Riemann surface $X_0(N)$ we have the corresponding differential $\omega_G := G(z)dz$. By the residue theorem,

$$\sum_{Q \in X_0(N)} \mathrm{Res}_Q \omega_G = 0,$$

where $\mathrm{Res}_Q \omega_G$ is the residue at the point Q with respect to the local coordinates on $X_0(N)$. Let $\mathrm{Res}_{z=\tau} G(z)$ be the residue of the function $G(z)$ at the point $\tau \in \mathcal{H}$. Let s be a cusp of $\Gamma_0(N)$ and let $\gamma\infty = s$, where $\gamma \in SL_2(\mathbb{Z})$. For $\tau \in \mathcal{H} \cup \{cusps\}$, let Q_τ be the point on $X_0(N)$ associated to τ under the usual identification. Let

$$f|_k \gamma(z) = \sum_{n \geq n_0} a_s(n) q_t^n, \quad t|N$$

be the expansion of f at the cusp s. Then

$$\mathrm{Res}_{Q_s} \omega_f = \frac{t}{2\pi i} a_s(0).$$

If $\tau \in \mathcal{H}$ and $l_{\tau,N}$ be as defined before, then

$$\mathrm{Res}_{Q_\tau} \omega_f = \frac{1}{l_{\tau,N}} \mathrm{Res}_\tau f.$$

Let

$$F(z) = \frac{\theta f}{f}(z) + \sum_{j=1}^{\sigma_0(N)} \beta_{j,N} E_2(t_j z),$$

where $\beta_{j,N}$ be as defined as in the Lemma. Then it follows that F is a modular form of weight 2 on $\Gamma_0(N)$ which vanishes at all the cusps, except at the cusp 0. Put

$$\omega_m = F(z) \cdot j_{m,N}(z) dz.$$

We first compute the residue of ω_m at the point Q_∞. Using the expression for $\frac{\theta f}{f}$ as given in Lemma 5, we see that

$$F(z) = h - \sum_{n=1}^{\infty} \sum_{d|n} c(d) d q^n + \sum_{j=1}^{\sigma_0(N)} \beta_{j,N} \left\{ 1 - 24 \sum_{n=1}^{\infty} \sigma(n/t_j) q^n \right\}.$$

Recall that

$$j_{m,N}(z) = q^{-m} + \sum_{n=0}^{\infty} c_{m,N}(n) q^n.$$

Therefore, the constant term in the expansion of $F(z) \cdot j_m(z)$ is equal to

$$-\sum_{d|m} c(d)d - 24 \sum_{j=1}^{\sigma_0(N)} \beta_{j,N}\sigma(m/t_j) + \left(h + \sum_{j=1}^{\sigma_0(N)} \beta_{j,N}\right) \cdot c_{m,N}(0).$$

Since by assumption $h + \sum_{j=1}^{\sigma_0(N)} \beta_{j,N} = 0$, we get

$$2\pi i \mathrm{Res}_{Q_\infty}\omega_m = -\sum_{d|m} c(d)d - 24 \sum_{j=1}^{\sigma_0(N)} \beta_{j,N}\sigma(m/t_j). \tag{7}$$

By construction, F is holomorphic at the cusp 0 and $j_m(z)$ has a simple zero at 0 and hence

$$\mathrm{Res}_{Q_0}\omega_m = 0. \tag{8}$$

Similarly, at the cusps $s \neq 0, \infty$, the function F has a zero and $j_m(z)$ is holomorphic and hence

$$\mathrm{Res}_{Q_s}\omega_m = 0 \text{ for all cusps } s \neq 0, \infty. \tag{9}$$

Let Q_τ be a point in $Y_0(N)$ corresponding to $\tau \in \mathcal{H}$. Since $E_2(z)$ and $j_m(z)$ are holomorphic functions on $\mathcal{H}$, for $\tau \in \mathcal{H}$,

$$\begin{aligned} 2\pi i \mathrm{Res}_{z=\tau} F(z) \cdot j_m(z) &= 2\pi i \left(\mathrm{Res}_{z=\tau} \frac{\theta f}{f}\right) \cdot j_m(\tau) \\ &= \mathrm{ord}_\tau(f) \cdot j_m(\tau). \end{aligned}$$

Therefore,

$$2\pi i \mathrm{Res}_{Q_\tau}\omega_m = \upsilon_{\tau,N}(f) \cdot j_m(\tau). \tag{10}$$

Using (7), (8), (9), (10), the fact that the residues sum to zero, and the bijection between points $\tau \in \mathcal{F}_N$ and points $Q_\tau \in Y_0(N)$, we conclude that

$$-\sum_{d|m} c(d)d - 24 \sum_{j=1}^{\sigma_0(N)} \beta_{j,N}\sigma(m/t_j) + \sum_{\tau \in \mathcal{F}_N} \upsilon_{\tau,N}(f) \cdot j_m(\tau) = 0.$$

In other words,

$$\sum_{\tau \in \mathcal{F}_N} \upsilon_{\tau,N}(f) \cdot j_m(\tau) = \sum_{d|m} c(d)d + 24 \sum_{j=1}^{\sigma_0(N)} \beta_{j,N}\sigma(m/t_j). \tag{11}$$

This completes the proof. □

Proof of Theorem 2. By Lemma 7,

$$\sum_{d|n} c(d)d = \sum_{\tau \in \mathcal{F}_N} \upsilon_{\tau,N}(f) \cdot j_{n,N}(\tau) - 24 \sum_{j=1}^{\sigma_0(N)} \beta_{j,N}\sigma(n/t_j).$$

Taking Möbius inversion, we get

$$n \cdot c(n) = \sum_{d|n} \mu(d) \sum_{\tau \in \mathcal{F}_N} \upsilon_{\tau,N}(f) \cdot j_{n/d,N}(\tau) - 24 \sum_{j=1}^{\sigma_0(N)} \beta_{j,N} \sum_{d|n} \mu(d)\sigma(n/dt_j).$$

Note that

$$\begin{aligned}\sum_{d|n} \mu(n/d) \sum_{j=1}^{\sigma_0(N)} \beta_{j,N}\sigma(d/t_j) &= \sum_{j=1}^{\sigma_0(N)} \beta_{j,N} \left(\sum_{d|n} \mu(n/d)\sigma(d/t_j)\right) \\ &= \sum_{j=1}^{\sigma_0(N)} \beta_{j,N} \frac{n}{t_j} \chi_{\mathbb{Z}}(n/t_j).\end{aligned}$$

Therefore,

$$c(n) = \frac{1}{n} \sum_{d|n} \mu\left(\frac{n}{d}\right) \sum_{\tau \in \mathcal{F}_N} \upsilon_{\tau,N}(f) j_{d,N}(\tau) - 24 \sum_{j=1}^{\sigma_0(N)} \frac{\beta_{j,N}}{t_j} \chi_{\mathbb{Z}}\left(\frac{n}{t_j}\right).$$

Proof of Theorem 3. Suppose that f is as in Theorem 3. Then by Lemma 7, Lemma 5 and Theorem 2, and (2), we have

$$\begin{aligned}\frac{\theta f}{f}(z) &= h - \sum_{n=1}^{\infty} \left(\sum_{d|n} c(d)d\right) q^n \\ &= h - \sum_{n=1}^{\infty} \sum_{\tau \in \mathcal{F}_N} \upsilon_{\tau,N}(f) j_{n,N}(\tau) q^n + 24 \sum_{n=1}^{\infty} \sum_{j=1}^{\sigma_0(N)} \beta_{j,N}\sigma(n/t_j) q^n \\ &= - \sum_{\tau \in \mathcal{F}_N} \upsilon_{\tau,N}(f) H_{\tau,N}(z) - \sum_{j=1}^{\sigma_0(N)} \beta_{j,N} E_2(t_j z),\end{aligned}$$

from which it follows that

$$\frac{\theta f}{f}(z) = -f_{\theta,N}(z) - \sum_{j=1}^{\sigma_0(N)} \beta_{j,N} E_2(t_j z).$$

This completes the proof of Theorem 3.

Proof of Theorem 1. Let

$$\Phi(z) = \frac{\theta \phi_N}{\phi_N}(z) + \sum_{j=1}^{\sigma_0(N)} \alpha_{j,N} E_2(t_j z).$$

It is easy to see that Φ is a meromorphic modular form of weight 2 on $\Gamma_0(N)$. Further, Φ is holomorphic at all the cusps and by assumption (by the choice of $\alpha_{j,N}$) it vanishes at all the cusps. Hence, it is a cusp form. For the values of $N \in \mathcal{G}$, the subgroup $\Gamma_0(N)$ has genus zero and so the space of cusp forms of weight 2 on $\Gamma_0(N)$ has zero dimension. Therefore, $\Phi(z) = 0$, which gives

$$\frac{\theta\phi_N}{\phi_N}(z) = -\sum_{j=1}^{\sigma_0(N)} \alpha_{j,N} E_2(t_j z).$$

Since $\theta(\phi_N(z) - \phi_N(\tau)) = \theta(\phi_N(z))$, we have

$$\frac{\theta\phi_N(z)}{\phi_N(z) - \phi_N(\tau)} = -\sum_{j=1}^{\sigma_0(N)} \alpha_{j,N} E_2(t_j z) \frac{\phi_N(z)}{\phi_N(z) - \phi_N(\tau)}$$

or

$$\frac{\theta(\phi_N(z) - \phi_N(\tau))}{\phi_N(z) - \phi_N(\tau)} = -\sum_{j=1}^{\sigma_0(N)} \alpha_{j,N} E_2(t_j z) \frac{\phi_N(z)}{\phi_N(z) - \phi_N(\tau)}. \tag{12}$$

Now we apply Theorem 3 with $f = \phi_N(z) - \phi_N(\tau)$. Since f is univalent, we have

$$f_{\theta,N}(z) = H_{\tau,N}(z).$$

Therefore,

$$\frac{\theta(\phi_N(z) - \phi_N(\tau))}{\phi_N(z) - \phi_N(\tau)} = -H_{\tau,N}(z) - \sum_{j=1}^{\sigma_0(N)} \alpha_{j,N} E_2(t_j z). \tag{13}$$

Comparing (12) and (13), we get

$$\begin{aligned} H_{\tau,N}(z) &= \left(\frac{\phi_N(z)}{\phi_N(z) - \phi_N(\tau)} - 1\right) \sum_{j=1}^{\sigma_0(N)} \alpha_{j,N} E_2(t_j z) \\ &= \frac{\phi_N(\tau)}{\phi_N(z) - \phi_N(\tau)} \sum_{j=1}^{\sigma_0(N)} \alpha_{j,N} E_2(t_j z). \end{aligned}$$

Proof of Theorem 1 is now complete.

Proof of Theorem 4. Define $b(m) = \sum_{d|m} c(d)d$. Then by Lemma 7, we have

$$b(m) = \sum_{\tau \in \mathcal{F}_N} v_{\tau,N}(f) j_{m,N}(\tau) - 24 \sum_{j=1}^{\sigma_0(N)} \beta_{j,N} \sigma\left(\frac{m}{t_j}\right). \tag{14}$$

Now by Lemma 5, we get

$$\left(q^h + \sum_{n=h+1}^{\infty} a(n) q^n\right)\left(h - \sum_{n=1}^{\infty} b(n) q^n\right) = hq^h + \sum_{n=h+1}^{\infty} a(n) n q^n.$$

Expanding the left hand side and comparing it to the right, we have for $n \geq 1$,

$$na(h+n) = -b(1)a(h+n-1) - b(2)a(h+n-2) - \cdots - b(n). \qquad (15)$$

The above recurrence relates the usual complete symmetric functions and power sums. Its solution is well known (see for instance [10]). Therefore, we get

$$b(n) = -na(h+n) + n \sum_{\substack{m_1+2m_2+\cdots+(n-1)m_{n-1}=n,\\ m_1,m_2,\ldots,m_{n-1}\geq 0}} (-1)^{m_1+\cdots+m_{n-1}} \times \frac{(m_1+\cdots+m_{n-1}-1)!}{m_1!\ldots m_{n-1}!} a(h+1)^{m_1}\ldots a(h+n-1)^{m_{n-1}}.$$

Comparing the above equation with (14), we get

$$\begin{aligned} a(h+n) = &\sum_{\substack{m_1+2m_2+\cdots+(n-1)m_{n-1}=n,\\ m_1,m_2,\ldots,m_{n-1}\geq 0}} (-1)^{m_1+\cdots+m_{n-1}} \frac{(m_1+\cdots+m_{n-1}-1)!}{m_1!\ldots m_{n-1}!} \\ &\times a(h+1)^{m_1}\ldots a(h+n-1)^{m_{n-1}} + \frac{24}{n}\sum_{i=1}^{\sigma_0(N)} \beta_{i,N}\sigma\left(\frac{n}{t_i}\right) \\ &- \frac{1}{n}\sum_{\tau\in\mathcal{F}_N} \upsilon_{\tau,N}(f) j_{n,N}(\tau) \\ = &F_n(\beta_{1,N},\ldots,\beta_{\sigma_0(N),N}, a(h+1), a(h+2),\ldots,a(h+n-1)) \\ &- \frac{1}{n}\sum_{\tau\in\mathcal{F}_N} \upsilon_{\tau,N}(f) j_{n,N}(\tau). \end{aligned}$$

Now, if we assume that f is holomorphic and non-vanishing on $\mathcal{H}$, then (14) reduces to

$$b(m) = -24 \sum_{j=1}^{\sigma_0(N)} \beta_{j,N}\sigma\left(\frac{m}{t_j}\right).$$

Substituting this directly in (15) we get (5). This completes the proof of Theorem 4.

Acknowledgment

We thank the referee for pointing out some minor corrections. Most of the work was carried out when the first author was a student at the Harish-Chandra Research Institute (HRI), Allahabad. She thanks HRI for all the support.

References

[1] S. Ahlgren, The theta-operator and the divisors of modular forms on genus zero subgroups, *Math. Research Letters*, **10(5–6)** (2003) 787–798.

[2] James R. Atkinson, Divisors of modular forms on $\Gamma_0(4)$, *J. Number Theory*, **112(1)** (2005) 189–204.

[3] R. E. Borcherds, Automorphic forms on $O_{s+2,2}(\mathbb{R})$ and infinite products, *Invent. Math.*, **120(1)** (1995) 161–213.

[4] Jan H. Bruinier, Winfried Kohnen and Ken Ono, The arithmetic of the values of modular functions and the divisors of modular forms, *Compositio Math.*, **140(3)** (2004) 552–566.

[5] D. Choi, On values of a modular form on $\Gamma_0(N)$, *Acta Arith.*, **121(4)** (2006) 299–311.

[6] S. Y. Choi, The values of modular functions and modular forms, *Canad. Math. Bull.*, **49(4)** (2006) 526–535.

[7] D. Dummit, H. Kisilevsky and J. McKay, Multiplicative products of η-functions. Finite groups—coming of age (Montreal, Que., 1982), 89–98, *Contemp. Math.*, **45**, *Amer. Math. Soc.*, Providence, RI (1985).

[8] Y. Choie and W. Kohnen, Special values of modular functions on Hecke groups, *Abh. Math. Sem. Univ. Hamburg*, **73** (2003) 289–295.

[9] W. Eholzer and N. Skoruppa, Product expansions of conformal characters, *Phys. Lett. B*, **388(1)** (1996) 82–89.

[10] I. G. Macdonald, Symmetric Functions and Hall Polynomials, Oxford University Press, Oxford (1995).

[11] K. Ono, The Web of Modularity: Arithmetic of the Coefficients of Modular Forms and q-series, *CBMS, Amer. Math. Soc.* (2004).

[12] B. Schoeneberg, Elliptic modular functions; an Introduction, Springer-Verlag (1974).

[13] G. Shimura, Introduction to the Arithmetic Theory of Automorphic Functions, Princeton University Press (1971).

[14] D. Zagier, Traces of singular moduli, in Motives, polylogarithms and Hodge theory, part I, Irvine, CA, 1998, *Int. Press Lecture Series*, vol. 3, I, eds F. Bogomolov and L. Katzarkov (International Press, Somerville, MA, 2002), 211–244.

TCV Proceedings No. 15, 2011, pp. 31–41.

Distribution Theorems for Saito-Kurokawa Lifts

Bernhard Heim

German University of Technology in Oman, Way No. 36, Building No. 331,
North Ghubrah, Muscat, Sultanate of Oman
e-mail: bernhard.heim@gutech.edu.om

Abstract. A Siegel modular cusp form F of weight k and degree 2 is a Saito-Kurokawa lift if and only if F satisfies the Maass relations. Recently it has been shown that this is equivalent to the requirement that F satisfies the Hecke duality relations. In this paper we prove that if the Hecke duality relations are fulfilled for all primes outside a set of primes with Dirichlet density smaller 1/8 then the cusp form is a Saito-Kurokawa lift. Further, for every weight k, there exists an effective constant c such that, if for all primes $p \leq c$ the Hecke duality relation is satisfied, then the form is a Saito-Kurokawa lift. Moreover we indicate further perspectives suggested by results of M. Manickam, B. Ramakrishnan and T. C. Vasudevan.

2000 Subject Classification: 11F

1. Introduction and statement of results

Let S_k^2 be the space of Siegel modular cusp forms of weight k and degree 2 with respect to the Siegel modular group $\Gamma^2 := Sp_2(\mathbb{Z})$. Let $A(T)$ be the Fourier coefficients of a cusp form $F \in S_k^2$, where $T = [n, r, m] = \begin{pmatrix} n & \frac{r}{2} \\ \frac{r}{2} & m \end{pmatrix}$ runs through the set of half-integral positive definite matrices T. Then F is a Saito-Kurokawa lift if and only if the Fourier coefficients satisfy the Maass relations

$$A[n, r, m] = \sum_{d \mid (n,r,m)} d^{k-1} \, A\left[\frac{nm}{d^2}, \frac{r}{d}, 1\right]. \tag{1.1}$$

Here d is summed over positive integers dividing the greatest common divisor of n, r and m, and we take the usual convention that $A(T) = 0$ for T not half-integral positive definite. See [3], [22], [10], [11] for more details and a summary. In [5], [6] we proved that F is a Saito-Kurokawa lift if and only if

$$A[n, r, pm] - A[np, r, m] = p^{k-1} \left(A\left[\frac{n}{p}, \frac{r}{p}, m\right] - A\left[n, \frac{r}{p}, \frac{m}{p}\right] \right) \tag{1.2}$$

for all $T = [n, r, m]$ and primes p. For a fixed prime p, the collection of such relations for all T we call the *p-Hecke duality* relation, since it reflects a duality property of Hecke operators for elliptic modular forms ([5], [4]). If F satisfies the p-Hecke

This work is partly supported by the *Max-Planck-Institut Mathematik Bonn* and *GUtech*, the German University of Technology in Oman.

duality relation for all primes p, say that F *satisfies the Hecke duality relations*. Employing a deep result of Chai–Faltings [2] on the Ramanujan conjecture for Siegel modular forms, Pitale and Schmidt [16] have recently extended this result. They have shown among other things that finitely many p-Hecke duality relations can be omitted. In this paper we extend these results in the Hecke duality aspect in two directions. First, we show that the p-Hecke duality conditions for a set of primes with Dirichlet density smaller than 1/8 can be omitted. Further, we show that, for fixed weight k, the p-Hecke duality relations need be checked only for finitely many primes p, which can be explicitly determined.

Theorem 1.1. *Let $\mathcal{R}$ be any set of primes with Dirichlet density smaller than 1/8. Let k be a positive even integer. Then we have the following property. For $F \in S_k^2$, F is a Saito-Kurokawa lift if and only if for all primes p outside $\mathcal{R}$ the p-Hecke duality relation (1.2) is satisfied.*

Finite sets have Dirichlet density 0. Hence this theorem contains the already mentioned results of [5] and [16].

Complementary to this result, we have:

Theorem 1.2. *For a positive even integer k there exists a constant $c(k)$, depending only upon k, such that $F \in S_k^2$ is a Saito-Kurokawa lift if and only if the p-Hecke duality relation is satisfied for all primes p with $p \leq c(k)$.*

The method used in this paper makes it possible to transfer multiplicity one theorems [18], [20], [21], [15] from elliptic modular forms to the space of Siegel modular forms of degree 2. Of course once this new method is established the main ingredient is in this case the beautiful and strong result of Ramakrishnan [20]. The second theorem indicates that there are sets $\mathcal{R}$ of density 1 (depending on k). The constant $c(k)$ is effective.

It is interesting to ask if the p-Hecke duality relation (1.2) for fixed prime number p itself, which contains infinitely many relations and is related to infinitely many Fourier coefficients, has an effective finiteness property. Of course since the vector space of Siegel modular forms of fixed weight k has a dimension asymptotic to k^3, we would somehow expect this, but it is not clear how to design such an algorithm in accordance to the p-Hecke duality relation. It is maybe interesting to note that the p-Hecke Duality relation of a cusp form F can be encoded in saying that F is in the kernel of an operator $|_k \bowtie T(p)$. Here $T(l)$ $(l \in \mathbb{N})$ is related to the Hecke operator of elliptic cusp forms. We prove

$$(F|_k \bowtie T(l)) = 0 \Leftrightarrow ((F|_k \bowtie T(l)))|_k \bowtie T(l) = 0. \tag{1.3}$$

Then we have

Theorem 1.3. *Let k, l be positive integers. Let k be even and $F \in S_k^2$. Then*

$$((F|_k \bowtie T(l)))|_k \bowtie T(l) \tag{1.4}$$

is a Siegel modular form with respect to a congruence subgroup

$$\Gamma^2(N) := \left\{ \gamma \in Sp(4)(\mathbb{Z}) | \gamma \equiv \begin{pmatrix} I_2 & 0 \\ 0 & I_2 \end{pmatrix} \; mod \; N \right\},$$

where the level $N = N(l)$ is effectively computable.

This has applications towards Siegel modular forms F and the functions $(F|_k \bowtie T(p))$. In this case we can choose the level to be equal to p^2.

Theorem 1.4. *Let k be a positive even integer and let p be a prime number. Let*

$$M := \left\lfloor \frac{k \cdot [\Gamma^2 : \Gamma^2(N(p))]}{10} \right\rfloor . \tag{1.5}$$

Let $F \in S_k^2$ and let $F|_k \bowtie T(p) \neq 0$. Then there exists a $\nu_0 \leq M$ such that the Taylor expansion is given by

$$(F|_k \bowtie T(p)) \begin{pmatrix} \tau & z \\ z & \tilde{\tau} \end{pmatrix} = \sum_{\nu=\nu_0}^{\infty} \chi_\nu^{F|_k \bowtie T(p)}(\tau, \tilde{\tau})\, z^{2\nu} \tag{1.6}$$

with $\chi_{\nu_0}^{F|_k \bowtie T(p)} \neq 0$.

From this we obtain by involving the differential operators $\mathcal{D}_{2\nu}$ (see (2.20) for a precise definition) the following application.

Corollary 1.5. *Let $F \in S_k^2$ be a Siegel modular form of even weight k and degree 2. Then F satisfies the p-Hecke duality relation if and only if for $\nu = 0, 1, \ldots, M$ the image of F with respect to $\mathcal{D}_{2\nu}$ is contained in the kernel of the operator $T(p) \otimes id - id \otimes T(p)$, i.e,*

$$(F|_k \bowtie T(p)) = 0 \Leftrightarrow \bigoplus_{\nu=0}^{M} \mathcal{D}_{2\nu}(F)|_{k+2\nu}(T(p) \otimes id - id \otimes T(p)) = 0. \tag{1.7}$$

Let $p_{k,\underline{2\nu}}(a, b)$ be the ultraspherical polynomial as defined in (2.19). Then

$$\mathcal{D}_{2\nu}(F)(\tau, \tilde{\tau}) = \sum_{n,m=1}^{\infty} \sum_{r, r^2 < 4nm} p_{k,\underline{2\nu}}(r, nm)\, A[n, r, m]\, e^{2\pi\, i n\tau} e^{2\pi\, i m\tilde{\tau}} \tag{1.8}$$

is an element of $\mathrm{Sym}^2(S_{k+2\nu})$. Let $(f_j)_j$ be the primitive Hecke eigenbasis of $S_{k+2\nu}$ then

$$\mathcal{D}_{2\nu}(F) = \sum_{i,j=1}^{\dim S_k} \alpha_{i,j}\, f_i \otimes f_j \quad (\alpha_{i,j} \in \mathbb{C}). \tag{1.9}$$

Let d_k be the dimension of S_k. Let

$$L(\tau, \tilde{\tau}) = \sum_{n,m=1}^{\infty} B(n, m)\, e^{2\pi\, i n\tau}\, e^{2\pi\, i m\tilde{\tau}} \in \mathrm{Sym}^2(S_{k+2\nu}).$$

Then $L = 0$ if for all $B(n, m) = 0$ for $n, m \leq d_{k+2\nu}$. Hence there are only finitely many conditions to check if $\mathcal{D}_{k,2\nu}(F)|_k \bowtie T(p) \equiv 0$. Combining this with the condition (1.8) demonstrates that the p-Hecke duality condition is actually a criteria with finitely many condition in praxis.

To generalize this work to Siegel modular forms with level structure seems to be a non trivial task. Nevertheless by the work of M. Manickam, B. Ramakrishnan and T. C. Vasudevan [13] which analyses the space of Saito-Kurokawa lifts inside the whole space of Siegel modular forms and also examines the micro structure (newforms, old forms) inside the Saito-Kurokawa space, there is hope to extend our knowledge on this very interesting subject, which has many applications in the field of modern number theory.

2. Proofs of the results

Siegel modular forms of degree 2 can be described in terms of tensor products of elliptic modular forms. This is one of the main results of [5]. In this paper we demonstrate that this viewpoint can be used to obtain strong results on the distribution of certain local properties, enabling us to decide whether or not a Siegel modular form is a Saito-Kurokawa lift. Hecke eigenforms which are not lifts are expected to satisfy the generalized Ramanujan conjecture.

We recall some notation and facts about modular forms. For a positive integer n, elements $\gamma = \left(\begin{smallmatrix} a & b \\ c & d \end{smallmatrix}\right)$ of the symplectic group $\mathrm{GSp}_n(\mathbb{R})$ with positive similitudes, act on the Siegel upper half-space $\mathfrak{H}_n$ of degree n by $\gamma(\tau) = (a\tau + b)(c\tau + d)^{-1}$. Here Sp_n denotes the symplectic group. For a positive integer k, denote by S_k^n the space of Siegel modular cusp forms of degree n and weight k. Elements are holomorphic functions F on $\mathfrak{H}_n$ which, for all $\gamma \in \Gamma_n := Sp_n(\mathbb{Z})$, satisfy the functional equation

$$F(\gamma(\tau)) = \det(c\tau + d)^k \, F(\tau). \tag{2.10}$$

Then F has a Fourier expansion

$$F(\tau) = \sum_{T \in \mathbb{A}_n} A(T) \, e^{2\pi i \, tr(T\tau)}, \tag{2.11}$$

where T runs over the set of all half-integral positive definite matrices. We let $|_k$ be the Petersson slash operator with the normalization:

$$F|_k\gamma(\tau) := \det(\gamma)^h \, \det(c\tau + d)^{-k} \, F(\gamma(\tau)), \tag{2.12}$$

where $\gamma \in \mathrm{GSp}_n(\mathbb{R})$ and $h = \frac{k}{2}$. We also could put $h = k - 1$. Then in the case $n = 1$ the eigenvalues and Fourier coefficients of primitive Hecke eigenforms coincide. But in this work we want to have the center of $\mathrm{GSp}_n(\mathbb{R})$ to act trivially. Then (2.10) is $F|_k\gamma = F$ for all $\gamma \in \Gamma_n$.

For convenience, we drop the index $n = 1$ in the case of elliptic modular forms. The vector space S_k is equipped with an action of *Hecke operators* T_m, indexed by non-negative integers m, defined by

$$T_m f(\tau) := m^{1-\frac{k}{2}} \sum_{n=1}^{\infty} \left(\sum_{d|(n,m)} d^{k-1} a\left(\frac{nm}{d^2}\right) \right) q^n, \text{ where } f(\tau) = \sum_{n=1}^{\infty} a(n) q^n. \tag{2.13}$$

with $q = e^{2\pi i\tau}$. These operators commute and are self-adjoint with respect to the Petersson inner product

$$\langle f, g\rangle := \int_{\Gamma\backslash\mathfrak{H}} f(\tau)\,\overline{g(\tau)}\,\mathrm{Im}(\tau)^{k-2}d\tau. \tag{2.14}$$

Hence, there exists a basis of simultaneous Hecke eigenforms. It is easy to see that Hecke eigenforms can be normalized such that the first Fourier coefficient is 1. Then the m-Fourier coefficient coincides with the m-th eigenvalue. A basis with this normalization property is denoted a basis of *primitive newforms*. The eigenvalues determine a newform uniquely. Since the algebra of Hecke operators at p is locally generated by the Hecke operators T_p for p prime, the associated eigenvalues $\lambda(p)$ of an eigenform uniquely determine the cusp form. This is the *multiplicity one* property of the space S_k.

A subset $\mathcal{R}$ of the set of primes is said to have Dirichlet density δ when the limit

$$\lim_{s\mapsto 1+}\frac{\sum_{p\in\mathcal{R}} p^{-s}}{\log\left(\frac{1}{s-1}\right)} \tag{2.15}$$

exists and is equal to δ. Then we ask the following question.

Let $f, g \in S_k$ two primitive newforms such that the eigenvalues for the operators T_p with p outside $\mathcal{R}$ are equal. For which sets $\mathcal{R}$ does this imply $f = g$?

Finite sets $\mathcal{R}$ have this property [14], [8]. This is the strong multiplicity one theorem. In this case the set $\mathcal{R}$ has Dirichlet density 0. But much more is known. Ramakrishnan [19] has proven that every set $\mathcal{R}$ with Dirichlet density at most 1/8 satisfies the strong multiplicity one condition. Certain improvements have also given by Rajan [17], [18]. Ramakrishnan's proof is based on analytic techniques. Rajan uses the Galois representation attached to a modular forms and invokes the Chebotarev density theorem.

On the other hand, given a weight k, there exists a constant c_k such that for all non-zero $f \in S_k$ there is at least one Fourier coefficient $a(m) \neq 0$ for $1 \leq m \leq c_k$. This follows from the finite-dimensionality of S_k. Non-trivial bounds are known (see for example [15]).

In the case of Siegel modular forms of degree 2 it is convenient to identify the index $T = \begin{pmatrix} n & \frac{r}{2} \\ \frac{r}{2} & m \end{pmatrix}$ for the Fourier coefficients with the quadratic form $[n, r, m]$.

2.1 *Proofs of theorem 1.1 and theorem 1.2*

As indicated in the introduction, $F \in S_k^2$ is a Saito-Kurokawa lift if and only if F satisfies the Maass relations (1.1). We have shown in [5] that this is equivalent to the requirement that F satisfies the infinitely many duality relations attached to the classical Hecke operators $T(l)$, $(l \in \mathbb{N})$ acting on elliptic modular forms. This can be further developed. Let $j : Sl_2(\mathbb{R}) \times Sl_2(\mathbb{R}) \hookrightarrow Sp_2(\mathbb{R})$ be the standard imbedding

$$\begin{pmatrix} a & b \\ c & d \end{pmatrix} \times \begin{pmatrix} a' & b' \\ c' & d' \end{pmatrix} \mapsto \begin{pmatrix} a & 0 & b & 0 \\ 0 & a' & 0 & b' \\ c & 0 & d & 0 \\ 0 & c' & 0 & d' \end{pmatrix}. \tag{2.16}$$

Identify j_1 and j_2 with the separate embeddings of $Sl_2(\mathbb{R})$ given by $j_1(\gamma) = j(\gamma \times I_2)$ and $j_2(\gamma) = j(I_2 \times \gamma)$. If $\gamma_1, \gamma_2 \in \Gamma^2 \in GSp_1(\mathbb{R})$ have the same factor of similitude then it also makes sense to consider $\gamma_1 \times \gamma_2$ as an element of $GSp_2(\mathbb{R})$. Let $\mathrm{Mat}(2, \mathbb{Z})_l$ be the set of all integral 2×2 matrices with determinant l. Then $\mathrm{Mat}(2, \mathbb{Z})_l$ is the union of finitely many left cosets $\Gamma \gamma_j$, where $\Gamma = Sp_1(\mathbb{Z})$. Let $\widetilde{\gamma_j} := l^{-\frac{1}{2}} \gamma_j$, and define, for $i = 1, 2$,

$$j_i(T(l))(F) := \sum_j F|_k j_i(\widetilde{\gamma_j}). \tag{2.17}$$

In [5] we have proven that $F \in S_k^2$ is a Saito-Kurokawa lift if and only if

$$j_1(T(l))(F) = j_2(T(l))(F) \qquad \text{for all } l \in \mathbb{N}. \tag{2.18}$$

Since the algebra of operators containing $T(l)$ is generated by $T(p)$, the property (2.18) is equivalent to the p-Hecke duality $j_1(T(p))(F) = j_2(T(p))(F)$ for all primes p. A straightforward calculation [6] translates this property into an identity among Fourier coefficients. We have $j_1(T(p)(F)) = j_2(T(p)(F))$ if and only if (1.2) is satisfied. Ultraspherical polynomials $p_{k,\underline{2\nu}}$ are defined as follows. Let $\nu \in \mathbb{N}_0$ and a and b be elements of a commutative ring. Then

$$p_{k,\underline{2\nu}}(a, b) := \sum_{\mu=0}^{\nu} (-1)^{\mu} \frac{(2\nu)!}{\mu!(2\nu - 2\mu)!} \frac{(k + 2\nu - \mu - 2)!}{(k + \nu - 2)!} a^{2\nu-2\mu} b^{\mu}. \tag{2.19}$$

Specializing the parameters, we have $p_{k,\underline{0}}(a, b) = 1$ and $p_{k,\underline{2\nu}}(0, 0) = 0$ for $\nu \in \mathbb{N}$. Write $(\tau, z, \tilde{\tau})$ for the point $\left(\begin{smallmatrix} \tau & z \\ z & \tilde{\tau} \end{smallmatrix}\right) \in \mathfrak{H}_2$. Define differential operators

$$\mathcal{D}_{k,\underline{2\nu}} F(\tau, \tilde{\tau}) := p_{k,\underline{2\nu}}\left(\frac{1}{2\pi i} \frac{\partial}{\partial z}, \left(\frac{1}{2\pi i}\right)^2 \frac{\partial}{\partial \tau} \frac{\partial}{\partial \tilde{\tau}} \right) F \Big|_{z=0} (\tau, \tilde{\tau}). \tag{2.20}$$

In the case $\nu = 0$ we get the pullback $F(\tau, 0, \tilde{\tau})$ of F on $\mathfrak{H} \times \mathfrak{H}$. Then we have the imbedding

$$\mathcal{D}_k := \bigoplus_{\nu=0}^{\lfloor \frac{k}{10} \rfloor} \mathcal{D}_{k,\underline{2\nu}} : S_k^2 \hookrightarrow \bigoplus_{\nu=0}^{\lfloor \frac{k}{10} \rfloor} \mathrm{Sym}^2(S_{k+2\nu}). \tag{2.21}$$

Here $\lfloor * \rfloor$ is the greatest integer less than or equal $*$. It follows from [5], that F is a Saito-Kurokawa lift if and only if for all $\nu \in \{0, 1, \ldots \lfloor \frac{k}{10} \rfloor\}$ and for all primes p we have

$$(T_p \otimes \mathrm{id})(\mathcal{D}_{k,\underline{2\nu}}(F)) = (\mathrm{id} \otimes T_p)(\mathcal{D}_{k,\underline{2\nu}}(F)). \tag{2.22}$$

For every ν let $(f_i^\nu)_i$ be a basis of primitive newforms in $S_{k+2\nu}$. Let $\lambda_i^\nu(p)$ be the eigenvalues of f_i^ν with respect to the Hecke operator T_p and $\mathcal{D}_{k,\underline{2\nu}}F = \sum_{i,j} \alpha_{i,j}^\nu\, f_i^\nu \otimes f_j^\nu$ with $\alpha_{i,j}^\nu \in \mathbb{C}$. Let $\mathcal{R}_{k,\nu} \subseteq \mathcal{P}$ be any subset of the set of primes of $\mathbb{Z}$ with the property that, if $\lambda_i^\nu(p) = \lambda_j^\nu(p)$ for all primes p outside $\mathcal{R}_{k,\nu}$, then $i = j$. This is for example fulfilled for the set given in [19].

If (2.22) is satisfied for all primes p outside $\mathcal{R}_{k,\nu}$, then

$$\mathcal{D}_{k,\underline{2\nu}}F = \sum_i \alpha_{i,i}^\nu\, f_i^\nu \otimes f_i^\nu. \tag{2.23}$$

Hence, F is a Saito-Kurokawa lift. Let $\mathcal{R}_k := \cap_{\nu=0}^{\lfloor \frac{k}{10} \rfloor} \mathcal{R}_{k,\nu}$, where $\mathcal{R}_{k,\nu}$ satisfies the properties just described. As noted in the introduction of section 2, every set $\mathcal{R}$ of primes with Dirichlet density at most 1/8 satisfies the properties of $\mathcal{R}_{k,\nu}$ (not depending on k and ν), hence we can also choose for $\mathcal{R}_k$ this $\mathcal{R}$. Hence we obtain: F is a Saito-Kurokawa lift if and only if $j_1(T(p))(F) = j_2(T(p))(F)$ for all primes outside $\mathcal{R}$. This proves Theorem 1.1. The eigenvalues of f_i^ν, f_j^ν are the same for all $0 \le \nu \le \lfloor \frac{k}{10} \rfloor$ and primes $2 \le p \le \dim S_{k+2\lfloor \frac{k}{10} \rfloor}$. Hence Theorem 1.2 is obvious.

2.2 *Properties of the Hecke duality operator*

Let $F \in S_k^2$ and let p be a prime number. Then the p-Hecke duality relation is given by

$$A[n,r,pm] - A[np,r,m] = p^{k-1}\left(A\left[\frac{n}{p},\frac{r}{p},m\right] - A\left[n,\frac{r}{p},\frac{m}{p}\right],\right)$$

where $T = [n,r,m]$ runs through all positive definite binary quadratic forms $nx^2 + rxy + my^2$. This relation is satisfied if and only if $j_1T(p)(F) = j_2T(p)(F)$. This can be put together to the following useful definition.

Definition. *Let $k,l \in \mathbb{N}$. Let F be a holomorphic function on $\mathfrak{H}$ which has the property $(F|_k j_1(\gamma_1))|_k j_2(\gamma_2) = F$ for all $\gamma_1,\gamma_2 \in \Gamma$. Then we define the **Hecke duality operator** $|_k \bowtie$ by*

$$F|_k \bowtie T(l) := j_1(T(l))(F) - j_2(T(l))(F). \tag{2.24}$$

By abuse of notation we say that $|_k \bowtie T(l)$ is an operator on the space of modular forms. Let $F \in S_k^2$. Then $F \in S_k^2$ satisfies the p-Hecke duality property if and only if F is in the kernel of $|_k \bowtie T(p)$. At this point we note that in general we can not assume that $F|_k \bowtie T(l)$ is a Siegel modular form for some congruence subgroup of Γ^2. This makes it somehow difficult to determine the vanishing order with respect to the variable z. Here we assume the standard parametrization of an element $Z = \left(\begin{smallmatrix} \tau & z \\ z & \tilde{\tau} \end{smallmatrix}\right) \in \mathfrak{H}_2$ in the Siegel upper half-space of degree 2. Nevertheless we have the following surprising result.

Proposition 2.6. *Let $k,l \in \mathbb{N}$. Let k be even and $F \in S_k^2$. Then*

$$F|_k \bowtie T(l) = 0 \Leftrightarrow (F|_k \bowtie T(l))\,|_k \bowtie T(l) = 0. \tag{2.25}$$

Proof. Note that the iteration of the operators $\bowtie T(l)$ is well-defined, since $H := F|_k \bowtie T(l)$ is invariant with respect to $|_k j_1(\tilde{\gamma})$ and $|_k j_2(\tilde{\gamma})$ for $\gamma \in \Gamma$.

Let G be a holomorphic function on the Siegel upper half-space of degree 2. Then $G \equiv 0$ if and only if $\mathcal{D}_{k,2\nu}(G) \equiv 0$ for one even positive integer k and all $\nu \in \mathbb{N}_0$. If $F \in S_k^2$ then we have

$$F|_k \bowtie T(l) \equiv 0 \Leftrightarrow \bigoplus_{\nu=0}^{\infty} \mathcal{D}_{k,2\nu}(F|_k \bowtie T(l)) \equiv 0$$

$$\Leftrightarrow \bigoplus_{\nu=0}^{\infty} \mathcal{D}_{k,2\nu}(F)|_{k+2\nu}(T(l) \otimes \mathrm{id} - \mathrm{id} \otimes T(l)) \equiv 0.$$

Let $(f_i^{(\nu)})_i$ be a Hecke eigenbasis of $S_{k+2\nu}$ with $T(l)$ eigenvalues $\lambda_i^{(\nu)}(l)$. Let $\mathcal{D}_{k,2\nu}(F) = \sum_{i,j=1}^{\dim S_{k+2\nu}} \alpha_{i,j}^{(\nu)} f_i^{(\nu)} \otimes f_j^{(\nu)}$. Then

$$F|_k \bowtie T(l) \equiv 0 \Leftrightarrow \bigoplus_{\nu=0}^{\infty} \sum_{i,j=1}^{\dim S_{k+2\nu}} \alpha_{i,j}^{(\nu)}(\lambda_i^{(\nu)}(l) - \lambda_j^{(\nu)}(l))\, f_i^{(\nu)} \otimes f_j^{(\nu)} \equiv 0$$

$$\Leftrightarrow \alpha_{i,j}^{(\nu)}(\lambda_i^{(\nu)}(l) - \lambda_j^{(\nu)}(l)) = 0 \quad \text{for all } \nu, i, j.$$

Now we put $H := F|_k \bowtie T(l)$. Then

$$H|_k \bowtie T(l) \equiv 0 \Leftrightarrow \bigoplus_{\nu=0}^{\infty} \mathcal{D}_{k,2\nu}(H)|_{k+2\nu}\, (T(l) \otimes \mathrm{id} - \mathrm{id} \otimes T(l)) \equiv 0$$

$$\Leftrightarrow \bigoplus_{\nu=0}^{\infty} \sum_{i,j=1}^{\dim S_{k+2\nu}} \alpha_{i,j}^{(\nu)}(\lambda_i^{(\nu)}(l) - \lambda_j^{(\nu)}(l))\, f_i^{(\nu)} \otimes f_j^{(\nu)}$$

$$|_{k+2\nu}\, (T(l) \otimes \mathrm{id} - \mathrm{id} \otimes T(l)) \equiv 0$$

$$\Leftrightarrow \bigoplus_{\nu=0}^{\infty} \sum_{i,j=1}^{\dim S_{k+2\nu}} \alpha_{i,j}^{(\nu)}(\lambda_i^{(\nu)}(l) - \lambda_j^{(\nu)}(l))^2\, f_i^{(\nu)} \otimes f_j^{(\nu)} \equiv 0$$

□

Once the tool of invertible differential operators [5] is available the proof is straightforward. The main point here is the existence of the property and some consequences.

One application one might be interested in is given by the following observation. Let $l \in \mathbb{N}$ be square-free and larger then one. Then it is not possible to embed $T(l)$ via j_1 or j_2 into $\mathrm{Sp}_2(\mathbb{Q})$ or even $\mathrm{GSp}_2(\mathbb{Q})$. This has been the reason why we had to take roots in our definition of $j_1(T(l)$ and $j_1(T(l))$. This has the consequence that we can not assume in general that $|_k \bowtie T(l)$ maps a Siegel modular form to a Siegel modular form (for some congruence subgroup).

Now let $F \in S_k^2$ and let p be a prime. Then we fix a disjoint Γ-left coset decomposition of $\Gamma\left(\begin{smallmatrix}1&0\\0&p\end{smallmatrix}\right)\Gamma$ with representatives $(\gamma_i)_i$. Then we have

$$\begin{aligned} H &:= (F|_k \bowtie T(p))|_k \bowtie T(p) \\ &= \sum_{i,j} F|_k j_1\left(\begin{pmatrix}p^{-1} & 0\\ 0 & p^{-1}\end{pmatrix}\gamma_i\gamma_j\right) + j_2\left(\begin{pmatrix}p^{-1} & 0\\ 0 & p^{-1}\end{pmatrix}\gamma_i\gamma_j\right) \\ &\quad - 2\sum_{i,j} F|_k(\gamma_i \times \gamma_j). \end{aligned} \tag{2.26}$$

But this shows that $H = \sum_t F|_k g_t$ is a finite sums with $g_t \in \mathrm{GSp}_2(\mathbb{Q})$. Each summand is a Siegel modular form with respect to $\Gamma^2(N_i)$, for $N_i \in \mathbb{N}$ suitable. We put $N = \max_i N_i$. Then H is a Siegel modular form with respect to $\Gamma^2(N)$. It is obvious that N can be effectively computed. Hence this proves Theorem 1.3.

2.3 *Vanishing orders*

Theorem 2.7. *Let k be a positive even integer. Let $F \in S_k^2(\Gamma^2(N))$ with Taylor expansion*

$$F\begin{pmatrix}\tau & z\\ z & \tilde{\tau}\end{pmatrix} = \sum_{\nu=0}^{\infty} \chi_{2\nu}(\tau,\tilde{\tau})\, z^{2\nu}. \tag{2.27}$$

Let $\mathcal{D}_{k,2\nu}(F) \in Sym^2(S_{k+2\nu}(\Gamma^2(N))$ be the associated modular forms. If

$$\mathcal{D}_{k,2\nu}^{\kappa}(F) := \oplus_{\nu=0}^{\kappa}\mathcal{D}_{k,2\nu}(F) = 0, \tag{2.28}$$

where $\kappa = \lfloor \frac{k\cdot\kappa(N)}{10} \rfloor$ with $\kappa(N)$ the index of $\Gamma^2(N)$ in Γ^2, then $F = 0$.

Proof. If $h \in \Gamma^2(N)$ then $F|_k h = F$ and if g_i runs through a representative system of $\Gamma^2(N)\backslash\Gamma^2$ then $F|_k g_i$ is well-defined. We put

$$G := \prod_{i=1}^{\kappa(N)} F|_k g_i \in M_{k\cdot\kappa(N)}^2(\Gamma^2). \tag{2.29}$$

In [5], section 2 we have proven the following: let $H \in S_k^2(\Gamma^2)$ and $m \in \mathbb{N}_0$ then

$$(\chi_{2\nu}^H(\tau,\tilde{\tau}))_\nu^m = 0 \text{ if and only if } (\mathcal{D}_{k,2\nu}(\tau,\tilde{\tau}))_{\nu=0}^m = 0. \tag{2.30}$$

Moreover let $m \geq \lfloor\frac{k}{10}\rfloor$ and let $(\chi_{2\nu}^H(\tau,\tilde{\tau}))_\nu^m = 0$ then $H = 0$. Hence we can deduce that $G = 0$ if and only if all the Taylor coefficients

$$(\chi_{2\nu}^G(\tau,\tilde{\tau})_{\nu=0}^{\lfloor\frac{k\cdot\kappa(N)}{10}\rfloor} = 0. \tag{2.31}$$

It is obvious that $G = 0$ if and only $F = 0$. Now we assume that (2.28) is satisfied and deduce that $G = 0$, which implies the claim of the theorem. Let $G_i := F|_k g_i$. Then

$$\chi_{2n}^G = \sum_{\substack{(n_i)_i\in\mathbb{N},\\ \sum_i n_i = n}} \prod_{i=1}^{\kappa(N)} \chi_{2n_i}^{G_i}.$$

Let $G_1 = F$ then $\chi_{2l}^{G_1} = 0$ for $0 \leq l \leq \lfloor \frac{k \cdot \kappa(N)}{10} \rfloor$. But this leads to $\chi_{2n}^{G} = 0$ for

$$0 \leq n \leq \left\lfloor \frac{k \cdot \kappa(N)}{10} \right\rfloor.$$

Hence we obtain $G = 0$. □

Proof of Theorem 1.4. Let $G := F|_k \bowtie T(p)$. We have shown that $G \neq 0$ if and only if $H := G|_k \bowtie T(p) \neq 0$. Moreover H is a Siegel modular form of degree 2 and weight k with respect to the congruence group $\Gamma^2(N(p))$. Therefore we can apply the result of Theorem 2.7. Since the vanishing order of G with respect to the variable z is not larger than the vanishing order of H with respect to the same variable we have proven the theorem.

Acknowledgements

This paper was partly written during the authors stay at the MPIM in Bonn (Germany) and at the GUtech in Muscat (Oman). It is a pleasure to thank the MPIM and GUtech for its excellent working atmosphere and financial support. The author thanks the referee for many helpful comments.

References

[1] K. Bringmann and B. Heim, Hecke duality and Jacobi forms, arXiv:0711.2532v2 [math.NT] 15.11.2007.
[2] Ch.-L. Chai and G. Faltings, Degeneration of Abelian Varieties, **22**, *Ergebnisse der Mathematik*, Berlin, Heidelberg, New York (1990).
[3] M. Eichler and D. Zagier, The theory of Jacobi forms, *Progress in Mathematics.*, **55**, Boston-Basel-Stuttgart, Birkhäuser (1985).
[4] P. Garrett and B. Heim, Hecke duality of Ikeda lifts, Preprint.
[5] B. Heim, On the Spezialschar of Maass, ArXiv: 0801.1804v1 [math.NT].
[6] B. Heim, A strong symmetry of Eisenstein series, ArXiv: 0801.18104v1 [math.NT].
[7] T. Ikeda, On the lifting of elliptic cusp forms to Siegel cusp forms of degree $2n$, *Ann. of Math.*, **154(3)** (2001) 641–681.
[8] H. Jacquet and J. Shalika, On Euler products and a classification of automorphic representations, I and II, *Amer. J. Math.*, **103** (1981) 499–558 and 777–815.
[9] N. Kurokawa, Examples of eigenvalues of Hecke operators on Siegel cusp forms of degree two, *Inventiones Math.*, **49** (1978) 149–165.
[10] H. Maass, Über eine Spezialschar von Modulformen zweiten Grades I., *Invent. Math.*, **52** (1979) 95–104.
[11] H. Maass, Über eine Spezialschar von Modulformen zweiten Grades II., *Invent. Math.*, **53** (1979) 249–253.
[12] H. Maass, Über eine Spezialschar von Modulformen zweiten Grades III., *Invent. Math.*, **53** (1979) 255–265.
[13] M. Manickam, B. Ramakrishnan and T. C. Vasudevan, On the Saito-Kurokawa descent for congruence subgroups, *Manuscripta Math.*, **81** (1993) 161–182.
[14] T. Miyawaki, On automorphic forms on $GL(2)$ and Hecke operators, *Ann. of Math.*, **94** (1971) 174–189.

[15] C. Moreno, Analytic proof of the strong multiplicity one theorem, *Amer. J. Math.*, **107(1)** (1985) 163–206.
[16] A. Pitale and R. Schmidt, Ramanujan-type results for Siegel cusp forms of degree 2, Homepapage Ralf Schmidt.
[17] C. S. Rajan, Refinement of strong multiplicity one for automorphic representations of $Gl(n)$, *Proc. Amer. Math. Soc.*, **128** (1998) 691–700.
[18] C. S. Rajan, On strong multiplicity one for l-adic representations, *IMRN*, **3** (1998) 161–172.
[19] D. Ramakrishnan, Refinement of the strong multiplicity one theorem for $GL(2)$, *Invent. math.*, **116** (1994) 645–649.
[20] D. Ramakrishnan, Modularity of the Rankin-Selberg L-series, and multiplicity one for $SL(2)$, *Ann. math.*, **152** (2000) 45–111.
[21] D. Ramakrishnan, Recovering representation form square, Appendix to: W. Duke and E. Kowalski, A problem of Linnik for elliptic curves and mean-value estimates for automorphic representations, *Invent. Math.*, **139(1)** (2000) 1–39.
[22] D. Zagier, Sur la conjecture de Saito-Kurokawa (d'après H. Maass, Sém. Delange-Pisot-Poitou 1979/1980, *Progress in Math.*, **12** (1980) 371–394.

TCV Proceedings No. 15, 2011, pp. 43–50.

Another Quaternionic Maaß Space

Aloys Krieg

Lehrstuhl A für Mathematik, RWTH Aachen University, D-52056 Aachen
e-mail: krieg@mathA.rwth-aachen.de

Abstract. In this paper we apply results of Freitag and Salvati Manni in order to determine the graded ring of quaternionic modular forms of degree 2 over an order of index 2 in the Hurwitz order. Moreover we can determine the associated Maaß space for a non-trivial character, which consists of skew-symmetric modular forms. Considering an appropriate submanifold we obtain non-trivial skew-symmetric Hermitian modular forms for an imaginary quadratic number field K with discriminant $d_K \equiv 0 \,(\text{mod } 4)$ whenever

$$-d_K = \alpha^2 + \beta^2 + \gamma^2 \quad \text{with } \alpha, \beta, \gamma \in \mathbb{N}.$$

Keywords. Modular forms, quaternions, graded ring, Maaß space.

Dedicated to Professor T. C. Vasudevan on the Occasion of his 60th Birthday

1. Introduction

We use the same notation as in [K3]. Therefore let

$$\mathcal{H}(\mathbb{H}) = \{Z = X + iY \in M_2(\mathbb{H}) \otimes_{\mathbb{R}} \mathbb{C};\ Z = X + iY,\ X = \bar{X}',\ Y = \bar{Y}' > 0\}$$

denote the *quaternionic half-space* of degree 2 (cf. [FH], [K1]) with the transpose as an exceptional automorphism

$$\tau : \mathcal{H}(\mathbb{H}) \to \mathcal{H}(\mathbb{H}), \quad Z \mapsto Z'.$$

Consider the Hurwitz order

$$\mathcal{O} = \mathbb{Z}\omega + \mathbb{Z}i_1 + \mathbb{Z}i_2 + \mathbb{Z}i_3, \quad \omega = \frac{1}{2}(1 + i_1 + i_2 + i_3),$$

and denote the ideal of even Hurwitz quaternions by

$$\wp = (1 + i_1)\mathcal{O} = \mathcal{O}(1 + i_1).$$

The modular group of degree 2 over the Hurwitz quaternions is given by

$$\mathrm{Sp}(2; \mathcal{O}) := \{M \in M_4(\mathcal{O});\ \bar{M}' J M = J\},$$

$$J = \begin{pmatrix} 0 & -I \\ I & 0 \end{pmatrix}, \quad I = I^{(2)} = \begin{pmatrix} 1 & 0 \\ 0 & 1 \end{pmatrix}.$$

The extended modular group $\Gamma(\mathcal{O})$ is the group of transformations $\mathcal{H}(\mathbb{H}) \to \mathcal{H}(\mathbb{H})$ generated by

$$Z \mapsto M\langle Z\rangle = (AZ+B)(CZ+D)^{-1},$$

$$M = \begin{pmatrix} A & B \\ C & D \end{pmatrix} \in \mathrm{Sp}(2;\mathcal{O}), \quad M = \rho I^{(4)}, \quad \rho = \frac{i_2 - i_3}{\sqrt{2}},$$

(cf. [FH], [K1], [K3]). The space $[\Gamma(\mathcal{O}), k]^{\mathrm{sym}}$, $k \in 2\mathbb{Z}$, of symmetric quaternionic modular forms consists of all holomorphic functions $f : \mathcal{H}(\mathbb{H}) \to \mathbb{C}$ satisfying $f \circ \tau = f$ as well as

$$f|_k M(Z) := (\det(CZ+D)^{\vee})^{-k/2} \cdot f(M\langle Z\rangle) = f(Z) \tag{1}$$

for all $M \in \mathrm{Sp}(2;\mathcal{O})$, $M = \rho I$, where "$\vee$" stands for the representation of quaternions in terms of complex 2×2 matrices (cf. [K1]). Examples are given by the Siegel Eisenstein series

$$E_k \in [\Gamma(\mathcal{O}), k]^{\mathrm{sym}} \quad \text{for even } k \geq 4$$

(cf. [K3]). Freitag and Hermann [FH] introduced 6 *quaternionic theta constants*

$$Y_1 = \theta_{\binom{0}{0}}, \quad Y_2 = \theta_{\binom{0}{1}}, \quad Y_3 = \theta_{\binom{1}{0}}, \quad Y_4 = \theta_{\binom{1}{1}}, \quad Y_5 = \theta_{\binom{1}{\omega}}, \quad Y_6 = \theta_{\binom{1}{\bar\omega}},$$

where

$$\theta_a(Z) = \sum_{g \in a + \wp^2} e^{\pi i \bar{g}' Z g}, \quad a \in \mathcal{O}^2, \quad Z \in \mathcal{H}(\mathbb{H}).$$

These theta constants are modular forms of weight 2 for the extended principal congruence subgroup of level $\wp$

$$Y_j \in [\Gamma[\wp], 2],$$

i.e.

$$Y_j|_2 M = Y_j \quad \text{for all } M \in \mathrm{Sp}(2;\mathcal{O}), \quad M \equiv I^{(4)} \pmod{\wp}, \quad M = \omega I^{(4)}$$

and

$$Y_j|_2 \rho I^{(4)} = Y_j \circ \tau.$$

The behavior of Y_j under τ was determined in [K4].

Freitag and Salvati Manni [FSM] proved that the graded ring of quaternionic modular forms $\oplus_{k \in 2\mathbb{Z}} [\Gamma[\wp], k]$ is the polynomial ring in the modular forms

$$Y_1, Y_2, Y_3, Y_4, Y_5, Y_6, E_6,$$

which are algebraically independent. Hence the subring $\oplus_{k \in 2\mathbb{Z}} [\Gamma(\mathcal{O}), k]^{\mathrm{sym}}$ also consists of polynomials in these modular forms (cf. [K3]). Note that the invariance group in the latter case is an integral orthogonal group $\mathcal{O}^+(S)$,

$$S = \begin{pmatrix} 0 & 1 \\ 1 & 0 \end{pmatrix} \perp \begin{pmatrix} 0 & 1 \\ 1 & 0 \end{pmatrix} \perp \begin{pmatrix} 2 & 1 & 1 & 1 \\ 1 & 2 & 0 & 0 \\ 1 & 0 & 2 & 0 \\ 1 & 0 & 0 & 2 \end{pmatrix}.$$

2. Quaternionic modular forms

In this section we apply the results from [FSM] in order to describe the graded ring of modular forms with respect to the orthogonal group $\mathcal{O}^+(S_0)$,

$$S_0 = \begin{pmatrix} 0 & 1 \\ 1 & 0 \end{pmatrix} \perp \begin{pmatrix} 0 & 1 \\ 1 & 0 \end{pmatrix} \perp 2I^{(4)}.$$

This means that we consider the (non-maximal) order

$$\Lambda = \mathbb{Z} + \mathbb{Z}i_1 + \mathbb{Z}i_2 + \mathbb{Z}i_3 \subset \mathcal{O}.$$

For $f \in [\Gamma(\Lambda), k] \subset [\Gamma(\mathcal{O}), k]$ we require the invariance conditions (1)

$$f|_k M = f \quad \text{for } M \in \mathrm{Sp}(2;\Lambda), \quad M = \omega I^{(4)}, \quad \rho I^{(4)}.$$

Note that $\Lambda/\wp = \mathbb{F}_2$ and

$$\mathrm{Sp}(2;\Lambda)/\{M \in \mathrm{Sp}(2;\mathcal{O});\, M \equiv I^{(4)} \ (\mathrm{mod}\ \wp)\} \cong \mathrm{Sp}(2;\mathbb{F}_2) \cong S_6. \tag{2}$$

We can determine the representation of this group on the space spanned by $Y_1, \dots, Y_6$ just as in [K3] and dermine the ring of invariant polynomials. Using a Magma calculation we obtain $g_k \in [\Gamma(\Lambda), k]^{\mathrm{sym}}$ by

$$\begin{aligned}
g_4 &= Y_1^2 + 3(Y_2^2 + Y_3^2 + Y_4^2 + Y_5^2 + Y_6^2),\\
g_6 &= Y_1^3 - 9Y_1(Y_2^2 + Y_3^2 + Y_4^2 - (Y_5 + Y_6)^2) + 54Y_2Y_3Y_4,\\
g_8 &= Y_1^4 + 6Y_1^2(Y_2^2 + Y_3^2 + Y_4^2) + 24Y_1Y_2Y_3Y_4 + 9(Y_2^4 + Y_3^4 + Y_4^4)\\
&\quad + 6(Y_2^2Y_3^2 + Y_2^2Y_4^4 + 2Y_2^2(Y_5 + Y_6)^2 + Y_3^2Y_4^2 + 2Y_3^2(Y_5 + Y_6)^2\\
&\quad + 2Y_4^2(Y_5 + Y_6)^2) + 72Y_5Y_6(Y_5^2 + Y_6^2),\\
g_{10} &= Y_1^5 - 3Y_1^3(Y_2^2 + Y_3^2 + Y_4^2 + Y_5^2 + Y_6^2) - 27Y_1(Y_2^4 + Y_3^4 + Y_4^4 + Y_5^4 + Y_6^4)\\
&\quad - 54Y_1(Y_2^2Y_3^2 + Y_2^2Y_4^2 + Y_2^2Y_5^2 + Y_2^2Y_6^2 + Y_3^2Y_4^2 + Y_3^2Y_5^2 + Y_3^2Y_6^2 + Y_4^2Y_5^2\\
&\quad + Y_4^2Y_6^2 + Y_5^2Y_6^2) + 648Y_2Y_3Y_4Y_5Y_6,\\
g_{12} &= Y_1^6 + 45Y_1^2(Y_2^4 + Y_3^4 + Y_4^4 + Y_5^4 + Y_6^4) + 1080Y_1Y_2Y_3Y_4Y_5Y_6\\
&\quad + 18(Y_2^6 + Y_3^6 + Y_4^6 + Y_5^6 + Y_6^6) + 270(Y_2^2Y_3^2Y_4^2 + Y_2^2Y_3^2Y_5^2\\
&\quad + Y_2^2Y_3^2Y_6^2 + Y_2^2Y_4^2Y_5^2 + Y_2^2Y_4^2Y_6^2 + Y_2^2Y_5^2Y_6^2 + Y_3^2Y_4^2Y_5^2\\
&\quad + Y_3^2Y_4^2Y_6^2 + Y_3^2Y_5^2Y_6^2 + Y_4^2Y_5^2Y_6^2).
\end{aligned}$$

Thus we obtain

Theorem 1. *The graded ring $\oplus_{k\in 2\mathbb{Z}}[\Gamma(\Lambda), k]^{\text{sym}}$ of symmetric quaternionic modular forms with respect to $\Gamma(\Lambda)$ is the polynomial ring in the modular forms*

$$(Y_5 - Y_6)^2, g_4, g_6, E_6, g_8, g_{10}, g_{12},$$

which are algebraically independent.

Remarks.

(a) Due to [K3] the modular forms g_4, E_6, g_{10}, g_{12} belong to $[\Gamma(\mathcal{O}), k]^{\text{sym}}$.

(b) g_4 is equal to the theta series

$$\Theta(\,\cdot\,, S_{\mathbb{H}}), \quad S_{\mathbb{H}} = \begin{pmatrix} 2 & 1+i_1 \\ 1-i_1 & 2 \end{pmatrix},$$

over the Hurwitz order (cf. [K4]).

(c) As easily verified there exists a theta series over Λ, namely

$$\Theta(Z, S) = \sum_{G\in M_2(\Lambda)} e^{\pi i\, \text{trace}(\bar{G}^{tr} SG\cdot Z)}, \quad Z \in \mathcal{H}(\mathbb{H}),$$

$$S = \begin{pmatrix} 2 & i_1+i_2+i_3 \\ -i_1-i_2-i_3 & 2 \end{pmatrix},$$

which belongs to $[\Gamma(\Lambda), 4]^{\text{sym}}$. Indeed one has

$$\Theta(\,\cdot\,, S) = g_4 - \frac{15}{4}(Y_5 - Y_6)^2.$$

3. The Maaß space

According to (2) there exists a non-trivial character χ on $\text{Sp}(2;\Lambda)$ which is an extension of the non-trivial character on $\text{Sp}(2;\mathbb{Z})$ (cf. [D], [KW]). It satisfies

$$\chi\begin{pmatrix} I & T \\ 0 & I \end{pmatrix} = (-1)^{n+m+r_0+r_1+r_2+r_3},$$

whenever

$$T = \begin{pmatrix} n & r \\ \bar{r} & m \end{pmatrix}, \quad r = r_0 + r_1 i_1 + r_2 i_2 + r_3 i_3 \in \Lambda, \quad n, m \in \mathbb{Z}.$$

Hence each $f \in [\text{Sp}(2;\Lambda), k, \chi]$ possesses a Fourier expansion of the form

$$f(Z) = \sum_{T\in\Lambda_2, T\geq 0} \alpha_f(T) e^{\pi i\, \text{trace}(TZ)},$$

where

$$\Lambda_2 = \left\{T = \begin{pmatrix} n & r \\ \bar{r} & m \end{pmatrix}; n, m \in 1 + 2\mathbb{Z}, r \in \omega + \Lambda\right\}.$$

Given $T \in \Lambda_2$ denote the content of T by

$$\varepsilon(T) := \max\left\{l \in \mathbb{N}; \frac{1}{l}T \in \Lambda_2\right\}.$$

In analogy with [M] and [K2] or [G] we define the *Maaß space* $\mathcal{M}_k(\Lambda, \chi)$ to consist of all $f \in [\mathrm{Sp}(2; \Lambda), k, \chi]$ such that each $T \in \Lambda_2$ satisfies

$$\alpha_f(T) = \sum_{0<d|\varepsilon(T)} d^{k-1} \cdot \alpha_f \begin{pmatrix} mn/d^2 & r/d \\ \bar{r}/d & 1 \end{pmatrix}.$$

Let

$$Z = \begin{pmatrix} z_0 & w \\ * & z_1 \end{pmatrix} \in \mathcal{H}(\mathbb{H}), \quad w = w_0 + w_1 i_1 + w_2 i_2 + w_3 i_3,$$

then f possesses a Fourier-Jacobi expansion of the form

$$f(Z) = \sum_{m \in \mathbb{N} \text{ odd}} \varphi_m(z_0, w) e^{\pi i m z_1},$$

where

$$\varphi_m(z_0, w) = \sum_{T = \begin{pmatrix} n & r \\ \bar{r} & m \end{pmatrix} \in \Lambda_2} \alpha_f(T) e^{\pi i (n z_0 + 2\langle r, w\rangle)},$$

and

$$\langle r, w\rangle = r_0 w_0 + r_1 w_1 + r_2 w_2 + r_3 w_3.$$

Just as in the classical case it follows that

$$f \mapsto \varphi_1$$

is an isomorphism between $\mathcal{M}_k(\Lambda, \chi)$ and the space of Jacobi forms of weight k, character χ and index 1, since one has

$$\varphi_m(z_0, w) = \sum_{ad=m} \sum_{b \bmod d} \frac{1}{m} \varphi_1 \left(\frac{az+b}{d}, aw \right).$$

It follows from [FH] and [K4] that

$$Y_5 - Y_6 \in \mathcal{M}_2(\Lambda, \chi)$$

with Fourier coefficients

$$\alpha(T) = -8\, sgn(r_0 r_1 r_2 r_3) \sigma_1(\varepsilon(T)).$$

The first Fourier Jacobi coefficient is equal to

$$\varphi_1(z_0, w) = -8(\vartheta_\omega(z_0, w) - \vartheta_{\bar{\omega}}(z_0, w)),$$

whenever

$$\vartheta_a(z_0, w) = \sum_{g \in a + \wp} e^{\pi i (g\bar{g} \cdot z_0 + 2\langle g, w\rangle)}.$$

For arbitrary $f \in \mathcal{M}_k(\Lambda, \chi)$ we have

$$\varphi_1(z_0, w) = f_0(z_0) \cdot (\vartheta_\omega(z_0, w) - \vartheta_{\bar{\omega}}(z_0, w)), \tag{3}$$

where

$$f_0(z_0) = \sum_{n=0}^{\infty} \alpha_f \begin{pmatrix} 2n+1 & \omega \\ \bar{\omega} & 1 \end{pmatrix} e^{2\pi i n z_0}.$$

Since the behavior of $\vartheta_\omega - \vartheta_{\bar{\omega}}$ under the Jacobi group is well-known or can be derived from $Y_5 - Y_6 \in \mathcal{M}_2(\Lambda, \chi)$, we obtain an isomorphism between the Maaß space and the space of entire modular forms of weight $k-2$ with respect to $\mathrm{SL}_2(\mathbb{Z})$.

Theorem 2. *Let $k \in \mathbb{N}$ be even. Then the mapping*

$$\mathcal{M}_k(\Lambda, \chi) \to [\mathrm{SL}_2(\mathbb{Z}), k-2], \quad f \mapsto f_0,$$

is an isomorphism.

Remarks.

(a) Each $f \in \mathcal{M}_k(\Lambda, \chi)$ is a modular form with respect to $\Gamma(\Lambda)$ as can be seen from the behavior of φ_1. We obtain

$$f|_2\, \omega I = f, \quad f|_2\, \rho I = f \circ \tau = -f.$$

It particular f is skew-symmetric.

(b) Due to the determination of zeros of $Y_5 - Y_6$ in [FH] we conclude that every modular form $f \in \mathcal{M}_k(\Lambda, \chi)$ possesses a representation

$$f = (Y_5 - Y_6) \cdot h, \quad h \in [\Gamma(\Lambda), k-2]^{\mathrm{sym}}.$$

4. Application to Hermitian modular forms

Just as in [K4] we consider the embeddings of Hermitian modular groups, where we quote [V] as a general reference. Due to the non-trivial character we have to restrict to even discriminants (cf. [D]). Let K be an imaginary quadratic number field of even discriminant $D_K = -4d$ with ring of integers

$$\mathcal{O}_K = \mathbb{Z} + \mathbb{Z}\sqrt{-d}.$$

Choose $\alpha, \beta, \gamma \in \mathbb{Z}$ such that

$$d = \alpha^2 + \beta^2 + \gamma^2, \quad j := \frac{1}{\sqrt{d}}(\alpha i_1 + \beta i_2 + \gamma i_3) \in \mathbb{H}, \quad j^2 = -1$$

and let $\sqrt{-d} := j\sqrt{d}$. Now the Hermitian modular group is

$$\Gamma(K) := \{M \in M_4(\mathcal{O}_K);\ \bar{M}' J M = J\} = \Gamma(\Lambda) \cap M_4(\mathcal{O}_K)$$

and the Hermitian half-space is given by

$$\mathcal{H}(\mathbb{C}) = \{Z = X + iY \in \mathcal{H}(\mathbb{H});\ X, Y \in M_2(\mathbb{R} + \mathbb{R}j)\}$$

as a submanifold of $\mathcal{H}(\mathbb{H})$. We may identify i and j in this case. Denote the attached Maaß space in the space of Hermitian modular forms by $\mathcal{M}_k(\Gamma(K), \chi)$.

Theorem 3. *Under the above assumption one has*

(a)

$$f|_{\mathcal{H}(\mathbb{C})} \in \mathcal{M}_k(\Gamma(K), \chi) \quad \textit{for} \ \ f \in \mathcal{M}_k(\Lambda, \chi).$$

(b)

$$f|_{\mathcal{H}(\mathbb{C})} \equiv 0 \quad \textit{if and only if} \quad \alpha\beta\gamma = 0.$$

(c) *The mapping*

$$\mathcal{M}_k(\Lambda, \chi) \to \mathcal{M}_k(\Gamma(K), \chi), \quad f \mapsto f|_{\mathcal{H}(\mathbb{C})},$$

is injective, whenever $\alpha\beta\gamma \neq 0$.

Proof.

(a) Confer [K2]. Note that a Jacobi form of weight k and index 1 is mapped onto a Hermitian Jacobi form of weight k and index 1 by the restriction.

(b) If one of the entries α, β, γ is 0 then $\mathcal{H}(\mathbb{C})$ is contained in a five dimensional submanifold, i.e. in

$$\mathcal{H}^* = \{Z \in \mathcal{H}(\mathbb{H});\ w_3 = 0\}$$

without restriction. As

$$i_3^{-1} Z' i_3 = Z \quad \text{for} \ \ Z \in \mathcal{H}^*$$

any skew-symmetric f satisfies

$$f(Z) = 0 \quad \text{for all} \ \ Z \in \mathcal{H}^*.$$

Hence let $\alpha, \beta, \gamma \in \mathbb{N}$ without restriction. Then it suffices to show that

$$\vartheta_\omega(z_0, w) - \vartheta_{\bar{\omega}}(z_0, w) \not\equiv 0$$

for $w \in \mathbb{C} + \mathbb{C}j$ by virtue of (3). One obtains a Fourier expansion

$$\vartheta_\omega(z_0, w) - \vartheta_{\bar{\omega}}(z_0, w) = \sum_{n \in \mathbb{N}} \sum_{t \in \mathcal{O}_K} A(n, t) e^{\pi i (n z_0 + \langle t/\sqrt{-d}, w \rangle)}.$$

We conclude

$$A(1, \alpha + \beta + \gamma + \sqrt{-d}) = -8.$$

(c) The kernel is equal to $\{0\}$ due to b). □

For all $K = \mathbb{Q}(\sqrt{-d})$ such that d is the sum of three positive squares we therefore have constructed a non-zero skew-symmetric Hermitian modular form of weight 2 and character χ by the restriction of $Y_5 - Y_6$. The example with smallest absolute discriminant is $\mathbb{Q}(\sqrt{-6})$. All these Hermitian modular forms of weight 2 vanish on

$$\left\{ Z = \begin{pmatrix} z_0 & w \\ * & z_1 \end{pmatrix} \in \mathcal{H}(\mathbb{C});\ w = w_0 + w_1 j,\ w_0 = 0 \quad \text{or} \ \ w_1 = 0 \right\}.$$

References

[D] T. Dern, Multiplikatorsysteme und Charaktere Hermitescher Modulgruppen, *Monatsh. Math.*, **126** (1998) 109–116.

[FH] E. Freitag and C. F. Hermann, Some Modular Varieties of Low Dimension, *Adv. Math.*, **152** (2000) 203–287.

[FSM] E. Freitag and R. Salvati Manni, Some modular varieties of low dimension II, *Adv. Math.*, **214** (2007) 135–147.

[G] V. Gritsenko, The Fourier-Jacobi functions of n, *J. Sov. Math.*, **53** (1991) 243–252.

[K1] A. Krieg, Modular Forms on Half-Spaces of Quaternions, Lect. Notes Math, **1143**, Springer-Verlag, Berlin-Heidelberg-New York (1985).

[K2] A. Krieg, The Maaß space for the non-trivial multiplier system over the Hurwitz quaternions, *Arch. Math.*, **70** (1998) 211–218.

[K3] A. Krieg, The graded ring of quaternionic modular forms of degree 2, *Math. Z.*, **251** (2005) 929–942.

[K4] A. Krieg, Theta series over the Hurwitz quaternions, *Int. J. Number Theory*, **6** (2010) 25–36.

[KW] A. Krieg and S. Walcher, Multiplier systems for the modular group on the 27-dimensional exceptional domain, *Comm. Algebra*, **26** (1998) 1409–1417.

[M] H. Maaß, Über ein Analogon der Vermutung von Saito-Kurokawa, *Invent. Math.*, **60** (1980) 85–104.

[V] T. C. Vasudevan, Some problems on hermitian modular forms, PhD thesis, Bombay (1978).

TCV Proceedings No. 15, 2011, pp. 51–62.

Distribution of Residues Modulo *p* II

F. Luca[1] and R. Thangadurai[2]

[1]Mathematical Institute, UNAM, Ap. Postal, 61-3 (Xangari), CP 58089, Morelia, Michoacán, Mexico
[2]Department of Mathematics, Harish-Chandra Research Institute, Chhatnag Road, Jhunsi, Allahabad 211 019, India
e-mail: fluca@matmor.unam.mx; thanga@hri.res.in

Abstract. In this article, we study a problem of the following nature. Given a natural number $N \geq 2$, does there exist a positive integer $p_0(N)$ such that for every prime $p \geq p_0(N)$, there is $x \in (\mathbb{Z}/p\mathbb{Z})^*$ with $x, x+1, \ldots, x+N-1$ are all quadratic residues (respectively, quadratic non-residues) modulo p? In 1928, Brauer [3] proved the existence of $p_0(N)$ for quadratic residues as well as quadratic non-residues mod p. In this article, we shall give an explicit bound for $p_0(N)$ for both the cases. Also, we study a related problem in this direction.

On the Occasion of 60th Birthday of Prof. T. C. Vasudevan

1. Introduction

For any prime number p, the distribution of residues modulo p has been of great interest to Number Theorists for many decades. The set of all non-zero residues modulo p can be divided into two classes, namely, the set of all quadratic residues (or squares) and quadratic non-residues (or non-squares) modulo p. In natural numbers, there are no consecutive squares as the difference of two consecutive squares is at least twice the least one. In the modulo p situation, one can expect a string of consecutive squares. In this article, we deal with the following question, first dealt by Brauer [3].

Question. For any given natural number $N \geq 2$, can we find an integer $p_0(N)$ such that for every prime $p \geq p_0(N)$, there exists an element $x \in (\mathbb{Z}/p\mathbb{Z})^*$ with $x, x+1, x+2, \ldots, x+N-1$ are all quadratic residues (respectively, quadratic non-residues) modulo p? If $p_0(N)$ exists, then can we find the explicit value?

In 1928, Brauer [3] answered the above question and proved the existence of $p_0(N)$ for quadratic residues and non-residues cases.

For a given prime p, the set of all non-residues modulo p can be, further, divided into two classes, namely, the set of all primitive roots (or generators of $(\mathbb{Z}/p\mathbb{Z})^*$) and non-residues which are not primitive roots modulo p.

In 1956, L. Carlitz [4] answered the above question for the set of all primitive roots modulo p and proved the existence of $p_0(N)$ in this case. This was independently proved by Szalay [24] and [25]. Recently, Gun *et al.* in [12], [13] and [18], answered the above question for the complementary case and gave an explicit value of $p_0(N)$ in that case.

It is worth to mention that Vegh [27], [28], [29] and [30] also studied similar related problems for case of primitive roots modulo p.

Another related problem along this direction was considered by D. H. Lehmer and E. Lehmer [17] as follows.

Definition. *Let $N \geq 2$ be an integer and p be a sufficiently large prime number. Define $r(N, p)$ (respectively $n(N, p)$) to be the least positive integer r such that*

$$r, r+1, \ldots, r+N-1$$

are all quadratic residues (respectively, quadratic non-residues) mod p.

Define

$$\Gamma(N) = \limsup_{p\to\infty} r(N, p); \qquad \gamma(N) = \liminf_{p\to\infty} r(N, p)$$

and

$$\Delta(N) = \limsup_{p\to\infty} n(N, p); \qquad \delta(N) = \liminf_{p\to\infty} n(N, p).$$

In [17], they proved that $\Gamma(2) = 9$ and $\Gamma(N) = \infty$ for all $N \geq 3$. In this article, while surveying these results, we prove the upper bounds for $p_0(N)$ for the case of quadratic residues and non-residues modulo p. Also, we discuss the values of $\Delta(N)$, $\gamma(N)$ and $\delta(N)$ for every $N \geq 2$.

2. Quadratic residues modulo *p*

We shall start with the following theorem.

Theorem 1. *Let $N \geq 2$ be a given integer and $p(N)$ denote the least prime number which is $> N$. Then there are infinitely many primes p which are* $\equiv 1 \pmod 4$ *such that*

$$1, 2, \ldots, p(N)-1, -p(N)+1, -p(N)+2, \ldots, -1$$

all are quadratic residues modulo p.

Remark. The idea of the proof this theorem lies in the paper [21] of S. S. Pillai.

Proof. Let $N \geq 2$ be a given integer. First we claim that *if $p = 4m+1$, then any divisor d of m is a quadratic residue modulo p.*

If a and b are quadratic residues mod p, then ab is a quadratic residue modulo p. Therefore, it is enough to prove the claim for any prime divisor of m.

Let q be a prime divisor of m. If $q = 2$, then $p \equiv 1 \pmod 8$. Therefore, by quadratic reciprocity law, we get,

$$\left(\frac{2}{p}\right) = (-1)^{(p^2-1)/8} = 1.$$

Let q be an odd prime. Then, by the quadratic reciprocity law, we have

$$\left(\frac{q}{p}\right) = \left(\frac{p}{q}\right) = \left(\frac{1}{q}\right) = 1,$$

since $q|m$ and hence $p - 1 \equiv 0 \pmod q$. Thus, the claim follows.

Consider the sequence of positive integers

$$S := 4(N!) + 1, 2(4(N!)) + 1, 3(4(N!)) + 1, \ldots, k(4(N!)) + 1, \ldots.$$

Then the Dirichlet Prime Number Theorem predicts that there are infinitely many prime numbers q in this sequence S. For these primes, by the above claim, we see that $1, 2, \ldots, N$ are all quadratic residues.

Also, note that -1 is a quadratic residues modulo these primes q, as $q \equiv 1 \pmod 4$. Therefore for these primes, $-1, -2, \ldots, -N$ are all quadratic residues modulo q.

Note that for any given N, the integers $N+1, N+2, \ldots, p(N)-1$ are all composed of primes that are less than or equal to N. This is because, by the definition of $p(N)$, there is no prime in between $N + 1$ to $p(N) - 1$. Hence, by the above observation, every divisor of $N + 1, N + 2, \ldots, p(N) - 1$ is a quadratic residue modulo q. Hence, $N + 1, N + 2, \ldots, p(N) - 1$ are all quadratic residues modulo q. Thus, the theorem follows. □

Remark. For any prime q satisfying Theorem 1, any quadratic non-residue modulo q lies between $p(N)$ and $-p(N)$ modulo q.

We give a new proof of the result of Brauer [3] with explicit value $p_0(N)$ as follows.

Theorem 2. *Let $N \geq 2$ be an integer. Then for every prime $p > \exp(2^{2^{2^{N^2+10}}})$, we can find $x, x + 1, x + 2, \ldots, x + N - 1$, for some $x \in (\mathbb{Z}/p\mathbb{Z})^*$, which are quadratic residues modulo p.*

The proof of Theorem 2 is an application of the celebrated Theorem of T. Gowers [10] which states as follows.

Theorem A (T. Gowers, [10]). *Let $M \geq 2$ be any integer and $0 < \delta < 1$. Then whenever $L \geq L(M, \delta) = \exp(\delta^{2^{2^{M+9}}})$, any subset $A \subset \{1, 2, \ldots, L\}$ with $|A| \geq \delta L$ contains an arithmetic progression of length M.*

Proof of Theorem 2. Let $N \geq 2$ be a given integer. Let p be any prime such that $p > \exp(2^{2^{2^{N^2+10}}})$. Let A be denote the set of all quadratic residues modulo p. Therefore, $|A| = \frac{p-1}{2}$. Put $L = p - 1$, $\delta = \frac{1}{2}$ and $M = N^2 + 1$ in Theorem A. Clearly, by the hypothesis, L satisfies the conditions of Theorem A and hence there exists an arithmetic progression

$$a, a + d, a + 2d, \ldots, a + N^2 d$$

of length $N^2 + 1$ in A. That means, $a, a + d, a + 2d, \ldots, a + N^2 d$ are all quadratic residues modulo p.

If d is a quadratic residue modulo p, then so is d^{-1}. Thus, we get

$$ad^{-1}, ad^{-1}+1, \dots, ad^{-1}+N^2$$

are all quadratic residues modulo p and we are done.

Suppose d is a quadratic non-residue modulo p. If there is $r \le N$ such that r is a quadratic non-residue modulo p, then rd is a quadratic residue modulo p and so is $(rd)^{-1}$. Hence, we have a sub arithmetic progression

$$a+rd, a+2rd, \dots, a+Nrd$$

whose terms are all quadratic residues modulo p with the difference rd which is also a quadratic residue modulo p. Therefore, we get,

$$a(rd)^{-1}+1, a(rd)^{-1}+2, \dots, a(rd)^{-1}+N$$

are all quadratic residue modulo p and we are done again.

If there is no $r \le N$ such that r is a quadratic non-residue modulo p, then $1, 2, \dots, N$ are quadratic residue modulo p and we have done. Thus the theorem follows. □

Recently, J. Tanti and R. Thangadurai [26] proved that $r(p, N) \le (N-2)^2 4^N$ for all $N \ge 3$.

Theorem 3 (Lehmer and Lehmer, [17]). $\Gamma(2) = 9$ *and* $\Gamma(N) = \infty$ *for all* $N \ge 3$. *Also,* $\gamma(N) = 1$ *for all* $N \ge 2$.

Proof. First we shall prove that $\Gamma(2) \le 9$. It is enough to prove that $r(2, p) \le 9$ for every prime $p \ge 11$. If 10 is a quadratic non-residue mod p, then either 2 or 5 are quadratic residue mod p. Hence $(1, 2)$ or $(4, 5)$ are pairs of quadratic residues mod p. If 10 is a quadratic residue mod p, then $(9, 10)$ is a pair of quadratic residue mod p. Also, this happens for all prime $p \ge 11$. Thus $\Gamma(2) \le 9$. To see the equality, it is enough to prove that $r(2, p) = 9$ for infinitely many primes p. That is to prove that 10 is a quadratic residue modulo p for infinitely many primes p. However, this is, indeed, true. For instance (for the reference, see Chapter 7 in [7]), the primes $p \equiv 1 \pmod{40}$ for which 10 is a quadratic residue mod p and by Dirichlet's Prime Number Theorem, we have infinitely many such primes. Hence, $\Gamma(2) = 9$ follows.

To prove $\Gamma(N) = \infty$, for all $N \ge 3$, it is enough to prove that $\Gamma(3) = \infty$, as if $\Gamma(M) = m < \infty$ for some $M > 3$, then it follows that $\Gamma(3) \le m$. To prove $\Gamma(3) = \infty$, it suffices to prove that for any given positive integer R, we have $r(3, p) \ge R$ for infinitely many primes p.

Let R be a given positive integer. Let $q_1, q_2, \dots, q_m$ be all the primes $q \le R$. By quadratic reciprocity law, we know that primes p for which q_i is a quadratic residue (respectively, quadratic non-residue) modulo p belong to the set (respectively, the complementary set) of some arithmetic progressions of common difference $4q_i$. List those primes p for which q_i is a quadratic residue modulo p and $q_i \equiv 1 \pmod 3$ and those primes p for which q_j is a quadratic non-residue modulo p and $q_i \equiv -1$

(mod 3). Combine the progressions of the first kind with those of the second kind. By Dirichlet's Prime Number Theorem, there are infinitely many primes p such that

$$\left(\frac{q}{p}\right) \equiv q \pmod 3 \qquad (q \neq 3, q \leq R).$$

Therefore, by the multiplicativity of the Legendre symbols, we conclude that

$$\left(\frac{m}{p}\right) \equiv m \pmod 3 \qquad (m \not\equiv 0 \pmod 3), m \leq R).$$

Among any three consecutive positive integers $\leq R$, there is an integer $m \equiv -1 \pmod 3$ and for which

$$\left(\frac{m}{p}\right) \equiv -1 \pmod 3 \Longrightarrow \left(\frac{m}{p}\right) = -1.$$

Thus, we get $r(3, p) \geq R$.

To see, $\gamma(N) = 1$ for all $N \geq 2$, we apply Theorem 1. By Theorem 1, we have infinitely many primes for which $1, 2, \ldots, N$ are all quadratic residues modulo these primes. Therefore, $r(N, p) = 1$ for infinitely many primes p. Hence $\gamma(N) = 1$ for all $N \geq 2$. □

3. Quadratic non-residues modulo *p*

The proof of Theorem 2, in general, doesn't work, if we replace the quadratic residues by quadratic non-residues. By Theorem 1, it is clear that for infinitely many primes $p \equiv 1 \pmod 4$, the first quadratic non-residue r is $\geq p(N) > N$ for any given $N \geq 2$. Hence the proof of Theorem 2 doesn't work in this case. However, it does work in some cases as follows.

Theorem 4. *Let $N \geq 2$ be an integer. Then for every prime p which is $\equiv \pm 3 \pmod 8$ and $p > \exp(2^{2^{2^{2N+10}}})$, we can find $x \in (\mathbb{Z}/p\mathbb{Z})^*$ such that $x, x+1, x+2, \ldots, x+N-1$ are all quadratic non-residues modulo p.*

Proof. Proceeding as in the proof of Theorem 2 for $p \equiv \pm 3 \pmod 8$ and A equal to the set of all quadratic non-residues mod p, we get an arithmetic progression of numbers

$$a, a+d, a+2d, \ldots, a+2Nd$$

each of which is quadratic non-residue modulo p.

If d is a quadratic residue modulo p, then so is d^{-1}. Hence, we get that

$$ad^{-1}, ad^{-1}+1, ad^{-1}+2, \ldots, ad^{-1}+N$$

are all quadratic non-residues modulo p.

Suppose d is a quadratic non-residue modulo p. When $p \equiv \pm 3 \pmod 8$, we know that 2 is a quadratic non-residue modulo p. Hence $2d$ is a quadratic non-residue modulo p. Thus, we get that the numbers

$$a(2d)^{-1}+1, a(2d)^{-1}+2, \ldots, a(2d)^{-1}+N$$

are all quadratic residues modulo p. Therefore, the result follows. □

To generalize the idea of the proof of Theorem 4, we need the following lemmas. Though the following lemma is well-known, for the sake of completeness, we include the proof here. To prove it, we need the following theorem.

Let $n > 1$ be an integer and m be an integer such that $1 \leq m \leq n$ and $(m, n) = 1$. Let $\pi(x, n, m)$ be denote the number of primes $p \leq x$ and $p \equiv m \pmod{n}$ and $\phi(n)$ denote the Euler Phi-function which counts the number of integers m with $1 \leq m \leq n$ and $(m, n) = 1$. Then Siegel–Walfisz theorem states as follows.

Siegel–Walfisz Theorem (see e.g., [22], Satz 4.8.3). *For any $A > 1$ and fixed integers $1 \leq m \leq n$, we have*

$$\pi(x, n, m) = \frac{\pi(x)}{\phi(n)} + O\left(\frac{x}{(\log x)^A}\right)$$

holds for all large enough x.

Lemma 5. *Let $n > 1$ be any integer which is not a perfect square of an integer. Then, for all large enough x, we have,*

$$\sum_{p \leq x} \left(\frac{n}{p}\right) = o(\pi(x)),$$

where $\pi(x)$ counts the number of primes up to x.

Proof. Define a map

$$\chi : (\mathbb{Z}/n\mathbb{Z})^* \longrightarrow \{\pm 1\}$$

by

$$\chi(m) = \left(\frac{n}{m}\right) \text{ for every } 1 \leq m \leq n, \ (m, n) = 1,$$

where $\left(\frac{n}{m}\right)$ is the Kronecker symbol. Note that when $m = 1$, we define $\chi(1) = 1$. By the multiplicativity of the Kronecker symbol, it is clear that χ is a character modulo n. Hence, by the orthogonality relation, we get

$$\sum_{\substack{1 \leq m \leq n \\ (m,n)=1}} \chi(m) = 0.$$

For simplicity, we define,

$$\sum_{m \pmod{\mathrm{n}}^*} := \sum_{\substack{1 \leq m \leq n \\ (m,n)=1}} .$$

Now, consider

$$\sum_{p \leq x} \left(\frac{n}{p}\right) = \sum_{\ell \pmod{\mathrm{n}}^*} \sum_{\substack{p \equiv \ell \pmod{\mathrm{n}} \\ p \leq x}} \left(\frac{n}{\ell}\right) = \sum_{\ell \pmod{\mathrm{n}}^*} \sum_{\substack{p \equiv \ell \pmod{\mathrm{n}} \\ p \leq x}} \chi(\ell).$$

By interchanging the order of summation, we get,

$$\sum_{p \le x} \left(\frac{n}{p}\right) = \sum_{\ell \text{ (mod n)}^*} \chi(\ell)\pi(x, n; \ell),$$

where $\pi(x, n, \ell)$ denotes the number of primes $p \equiv \ell$ (modn) and $p \le x$. Walfisz's Theorem implies that for any fixed integer $A > 1$, we have

$$\pi(x, n, \ell) = \frac{\pi(x)}{\phi(n)} + O\left(\frac{x}{(\log x)^A}\right)$$

for every large enough x. Therefore, we get,

$$\sum_{p \le x} \left(\frac{n}{p}\right) = \frac{\pi(x)}{\phi(n)} \sum_{\ell \text{ (mod n)}^*} \chi(\ell) + O\left(\frac{\phi(n)x}{(\log x)^A}\right).$$

By the orthogonality relation, we, further, get,

$$\sum_{p \le x} \left(\frac{n}{p}\right) = O\left(\frac{\phi(n)x}{(\log x)^A}\right) = o(\pi(x)).$$

Hence, the lemma. □

Corollary 6. *For any integer $s \ge 2$ which is not a perfect square of an integer, then there are infinitely many primes p for which s is a quadratic non-residue modulo p.*

Proof. If there are only finitely many primes, say, $p_1, p_2, \ldots, p_r$ for which s is a quadratic non-residue, then for any $x > p_r$

$$\sum_{\substack{p \le x \\ p \ne p_i}} \left(\frac{s}{p}\right) = \pi(x) - r \ne o(\pi(x))$$

a contradiction to Lemma 5. Hence, there are infinitely many primes p for which s is a quadratic non-residue modulo p. □

Remark. Since 3 is a quadratic non-residue modulo p for every prime $p \equiv \pm 5$ (mod 12), we see that *for every prime $p \equiv \pm 5$ (mod 12) and $p > \exp(2^{2^{2^{3N+10}}})$, we can find $x \in (\mathbb{Z}/p\mathbb{Z})^*$ such that $x, x+1, \ldots, x+N-1$ are quadratic non-residue modulo these primes.* More generally, let $f(N)$ denotes an increasing function of N. Then we can find infinitely many primes satisfying $p > \exp(2^{2^{2^{sf(N)+10}}})$ where $2 \le s \le f(N)$ is not a perfect square of an integer and s is a quadratic non-residue for these primes (by Corollary 6). Then these primes satisfy the conclusion of Theorem 3.

Recently, J. Tanti and R. Thangadurai [26] proved that $n(p, N) \le (N-2)^2 4^N$ for all $N \ge 3$.

Theorem 7. $\Delta(N) = \infty$ *for all* $N \ge 2$.

Proof. Theorem 1 implies that there is a sequence of primes $p_1, p_2, \ldots, p_r, \ldots$, for which $n(N, p_i) \geq N$ for all i. Therefore $\Delta(N) \geq p(N)$ (the smallest prime $p > N$) for all $N \geq 2$. However, the least quadratic non-residue modulo p (denoted by $g(p)$) satisfies $g(p) \geq (\log p)(\log\log\log p)$ for infinitely many primes p (this result is due to Graham and Ringrose [9]). Therefore, $n(N, p) \geq (\log p)(\log\log p)$ for infinitely many primes p and consequently, $\Delta(N) = \infty$. □

Regarding $\delta(N)$, first we prove that $\delta(2) = 2$. For that we need to prove 2 and 3 are quadratic non-residues modulo p for infinitely many primes p. In [11], Gupta and Murty proved, using sieve theory, that

$$\#\left\{p \leq x : p - 1 = 2q \text{ or } 2q_1q_2, \left(\frac{2}{p}\right) = \left(\frac{3}{p}\right) = -1\right\} \geq \frac{cx}{\log^2 x}$$

for some $c > 0$. Therefore, by taking $x \to \infty$, we get there are infinitely many primes p for which 2, 3 are quadratic non-residues mod p. Thus, $\delta(2) = 2$ follows.

When $N = 3$, we can prove that $\delta(3) = 5$. Clearly, $\delta(3) \geq 5$, because 1 and 4 are perfect squares. For the upper bound, we need to prove that 5, 6, 7 are quadratic non-residues modulo p for infinitely many primes p. This has been achieved in [2]. Hence, $\delta(3) = 5$.

In the case of primitive roots modulo p, as we mentioned in the introduction, Carlitz [4] and Szalay [24] and [25] proved the existence of $p_0(N)$. Recently, Tanti and Thangadurai [26] proved that $p_0(N) \leq \exp(2^{2.77N})$ (for the primitive root case). In [13] and [18], we proved the existence of $p_0(N)$ for the case of non-residues which are not the primitive roots modulo p. Also we proved an upper bound for $p_0(N)$. In fact, in [18], we proved the following result which improves the result in [13].

Theorem B. *Let $\varepsilon \in (0, 1/2)$ be fixed and let $N \geq 2$ be an integer. If*

$$p \geq \max\{N^2(4/\varepsilon)^{2N}, N^{651N\log\log(10N)}\}$$

is a prime satisfying

$$\frac{\phi(p-1)}{p-1} \leq \frac{1}{2} - \varepsilon,$$

then there are N consecutive integers $n, \ldots, n+N-1$ that are quadratic non-residues but not primitive roots modulo p.

It is also possible to give a bound similar to Theorem B for $p_0(N)$ for the primitive root mod p case.

In 1976, Hausman [14] proved the existence of p_0 such that for every prime $p \geq p_0$, there exists an integer $g \leq p - 1$ and $(g, p - 1) = 1$ such that g is a primitive root modulo p. Recently, R. Thangadurai [26] proved that $p_0 \leq e^{110.8} \sim 1.318 \times 10^{48}$.

4. A related problem

Another related question is as follows. For a given non-empty subset $S = \{a_1, a_2, \ldots, a_\ell\}$ of $\mathbb{Z}$, can we find infinitely many primes p such that every element of S is a quadratic residue (respectively, non-residue) modulo p? If yes, what is the density of such primes for a given subset S?

In 1968, M. Fried [8] answered that there are infinitely many primes p for which a is a quadratic residue modulo p for every $a \in S$. Also, he provided a necessary and sufficient condition for a to be a quadratic non-residue modulo p for every $a \in S$. More recently, S. Wright [31] and [32] also studied this qualitative problem.

For a given prime p, the set of all quadratic non-residue modulo p is a disjoint union of the set of all generators g of $(\mathbb{Z}/p\mathbb{Z})^*$ (which are called primitive roots modulo p) and the complementary set containing all the non-residues which are not primitive roots modulo p.

A set P of prime numbers is said to have the *relative density* ε with $0 \leq \varepsilon \leq 1$, if

$$\varepsilon = \lim_{x\to\infty} \frac{|P \cap [1, x]|}{\pi(x)}$$

exits. Also, the following numbers count some special subsets of S.

(i) Let α_S denote the number of subsets T of S, including the empty one, such that $|T|$ is even and $\prod_{s\in T} s = m^2$ for some integer m; hence, $\alpha_S \geq 1$ for every S.
(ii) Let β_S denote the number of subsets T of S such that $|T|$ is odd and $\prod_{s\in T} s = m^2$ for some integer m.

Then the following theorems were proved by R. Balasubramanian, F. Luca and R. Thangadurai [2].

Theorem 8 ([2], 2010). *The relative density of the set of prime numbers p for which a is a quadratic residue modulo p for every $a \in S$ is*

$$\frac{\alpha_S + \beta_S}{2^\ell}.$$

Theorem 9 ([2], 2010). *We have, $\beta_S = 0$ if and only if the density of the set of primes p for which a is a quadratic non-residue modulo p for every $a \in S$ is*

$$\frac{\alpha_S}{2^\ell}.$$

We shall present the proof of Theorem 8.

Proof of Theorem 8. Let $\mathcal{P}(S)$ be the set of all distinct prime factors of $a_1 a_2 \cdots a_\ell$. Clearly, $|\mathcal{P}(S)|$ is finite. Let $x > 1$ be a real number. Consider the following counting function

$$S_x = \frac{1}{2^\ell} \sum_{\substack{p\leq x \\ p\notin\mathcal{P}(S)}} \left(1 + \left(\frac{a_1}{p}\right)\right) \cdots \left(1 + \left(\frac{a_\ell}{p}\right)\right).$$

Since the Legendre symbol is completely multiplicative, $\left(\frac{a_i}{p}\right)\left(\frac{a_j}{p}\right) = \left(\frac{a_i a_j}{p}\right)$, we see that

$$S_x = \frac{1}{2^\ell} \sum_{\substack{p\leq x \\ p\notin\mathcal{P}(S)}} \sum_{\substack{0\leq b_i\leq 1 \\ n=a_1^{b_1}\cdots a_\ell^{b_\ell}}} \left(\frac{n}{p}\right) = \sum_{\substack{0\leq b_i\leq 1 \\ n=a_1^{b_1}\cdots a_\ell^{b_\ell}}} \frac{1}{2^\ell} \sum_{\substack{p\leq x \\ p\notin\mathcal{P}(S)}} \left(\frac{n}{p}\right).$$

Note that if n is a perfect square, then $\left(\frac{n}{p}\right) = 1$ for each $p \notin \mathcal{P}(S)$. Thus, for these $\alpha_S + \beta_S$ values of n, the inner sum is

$$\frac{1}{2^\ell} \sum_{\substack{p \leq x \\ p \notin \mathcal{P}(S)}} \left(\frac{n}{p}\right) = \frac{1}{2^\ell}(\pi(x) - |\mathcal{P}(S)|).$$

For the remaining values of n (i.e., when n is not a perfect square), we apply Lemma 5 to get

$$\frac{1}{2^\ell} \sum_{\substack{p \leq x \\ p \notin \mathcal{P}(S)}} \left(\frac{n}{p}\right) = o(\pi(x)) \qquad \text{as} \quad x \to \infty.$$

Therefore,

$$S_x = \frac{1}{2^\ell}(\alpha_S + \beta_S)(\pi(x) - |\mathcal{P}(S)|) + o(\pi(x)),$$

and hence

$$\frac{S_x}{\pi(x)} = \frac{\alpha_S + \beta_S}{2^\ell}\left(1 - \frac{|\mathcal{P}(S)|}{\pi(x)}\right) + o(1).$$

Since $|\mathcal{P}(S)|$ is a finite number and it is elementary to see that as $x \to \infty$, $\pi(x) \to \infty$, we get

$$\lim_{x \to \infty} \frac{S_x}{\pi(x)} = \frac{\alpha_S + \beta_S}{2^\ell}.$$

This completes the proof of Theorem 8. □

This can be applied to the quadratic non-residue case as well. Take

$$S_x = \frac{1}{2^\ell} \sum_{\substack{p \leq x \\ p \notin \mathcal{P}(S)}} \left(1 - \left(\frac{a_1}{p}\right)\right) \cdots \left(1 - \left(\frac{a_\ell}{p}\right)\right),$$

and proceed as in the proof of Theorem 8. This yields Theorem 9.

For a given prime p, the set of all quadratic non-residue modulo p is a disjoint union of the set of all generators g of $(\mathbb{Z}/p\mathbb{Z})^*$ (which are called primitive roots modulo p) and the complement set contains all the non-residues which are not primitive roots modulo p.

In 1927, E. Artin [1] conjectured the following:

Artin's primitive root conjecture. Let $g \neq \pm 1$ be a square-free integer. Then there are infinitely many primes p such that g is a primitive root modulo p.

Note that it is not even known that for a given square-free integer, $g \neq \pm 1$, there exists a prime p such that g is a primitive root modulo p. The above Artin's conjecture asks for the existence infinitely many such primes. In 1967, Hooley [16] proved this conjecture assuming the (as yet) unresolved genearlized Riemann hypothesis for Dedekind zeta functions of certain number fields. In 1983, R. Gupta and M. R. Murty

[11] made the first breakthrough by showing the following: given three prime numbers a, b, c, then at least one of the thirteen numbers

$$\{ac^2, a^3b^2, a^2b, b^3c, b^2c, a^2c^3, ab^3, a^3bc^2, bc^3, a^2b^3c, a^3c, ab^2c^3, abc\}$$

is a primitive root modulo p for infinitely many primes p. Then later Heath-Brown [15] proved that among $\{a, b, c\}$ one is primitive root modulo p for infinitely many primes p. Similarly, using the method of Hooley, in 1976, K. R. Matthews [19] found a necessary and sufficient condition for a to be primitive root modulo p for every $a \in S$, under an unproved hypothesis.

The analogous question for a non-residue which is not a primitive root modulo a prime is relatively easier to handle. For example, in [20] it is proved that for a given g which is not a perfect square of an integer, there are infinitely many primes p for which g is a quadratic non-residue but not a primitive root modulo p, using the arithmetic of certain number fields.

Acknowledgement

We are thankful to the referee for carefully going through the previous version and for his/her suggestions to improve the presentation.

References

[1] E. Artin, Collected Papers, Addison-Wesley (1965).

[2] R. Balasubramanian, F. Luca and R. Thangadurai, On the exact degree of $\mathbb{Q}(\sqrt{a_1}, \sqrt{a_2}, \dots, \sqrt{a_\ell})$ over $\mathbb{Q}$, *Proc. Amer. Math. Soc.*, **138** (2010) 2283–2288.

[3] A. Brauer, Űber Sequenzen von Potenzresten, *Sitzungsberichte der Preubischen Akademie der Wissenschaften* (1928) 9–16.

[4] L. Carlitz, Sets of primitive roots, *Compositio Math.*, **13** (1956) 65–70.

[5] H. Davenport, On the distribution of quadratic residues (mod p), *J. London Math. Soc.*, **6** (1931) 49–54.

[6] H. Davenport, On the distribution of quadratic residues (mod p), *J. London Math. Soc.*, **8** (1933) 46–52.

[7] J. Esmonde and M. Ram Murty, Problems in Algebraic Number Theory, Graduate Texts in Mathematics, 190. Springer-Verlag, New York (1999).

[8] M. Fried, Arithmetical properties of value sets of polynomials, *Acta Arith.*, **15** (1968/69) 91–115.

[9] S. W. Graham and C. J. Ringrose, Lower bounds for least quadratic non residues. Analytic number theory (Allerton Park, IL, 1989), 269–309, *Progr. Math.*, 85, Birkhäuser Boston, Boston, MA (1990).

[10] W. T. Gowers, A new proof of Szemerédi's theorem, *Geom. Funct. Anal.*, **11** (2001) 465–588.

[11] R. Gupta and M. Ram Murty, A remark on Artin's conjecture, *Invent. Math.*, **78** (1984) 127–130.

[12] S. Gun, B. Ramakrishnan, B. Sahu and R. Thangadurai, Distribution of Quadratic non-residues which are not primitive roots, *Math. Bohem.*, **130** (2005) 387–396.

[13] S. Gun, F. Luca, P. Rath, B. Sahu and R. Thangadurai, Distribution of residues modulo p, *Acta Arith.*, **129** (2007) 325–333.

[14] M. Hausman, Primitive roots satisfying a coprime condition, *Amer. Math. Monthly*, **83** (1976) 720–723.
[15] D. R. Heath-Brown, Artin's conjecture for primitive roots, *Quart. J. Math. Oxford*, (2) **37** (1986) 27–38.
[16] C. Hooley, On Artin's conjecture, *J. Reine. Angew. Math.*, **225** (1967) 209–220.
[17] D. H. Lehmer and E. Lehmer, On runs of residues, *Proc. Amer. Math. Soc.*, **13** (1962) 102–106
[18] F. Luca, I. E. Shparlinski and R. Thangadurai, Quadratic non-residue verses primitive roots modulo p, *J. Ramanujan Math. Soc.*, **23** (2008) 97–104.
[19] K. R. Matthews, A generalisation of Artin's conjecture for primitive roots, *Acta Arith.*, **29** (1976) 113–146.
[20] P. Moree and R. Thangadurai, An analogue of Artin primitive root conjecture, Preprint (2007).
[21] S. S. Pillai, On the divisors of $a^n + 1$, *J. Indian Math. Soc.* (N. S), **6** (1942) 120–121.
[22] K. Prachar, Primzahlverteilung, Springer, New York (1957).
[23] S. Sen Gupta, Artin's conjecture; Unconditional approach and elliptic analogue, Master's thesis, Waterloo, Ontario, Canada (2008).
[24] M. Szalay, On the distribution of the primitive roots mod p (in Hungarian), *Mat. Lapok*, **21** (1970) 357–362.
[25] M. Szalay, On the distribution of the primitive roots of a prime, *J. Number Theory*, **7** (1975) 183–188.
[26] J. Tanti and R. Thangadurai, Distribution of residues and primitive roots, Preprint (2011).
[27] E. Vegh, Primitive roots modulo a prime as consecutive terms of an arithmetic progression, *J. Reine Angew. Math.*, **235** (1969) 185–188.
[28] E. Vegh, Primitive roots modulo a prime as consecutive terms of an arithmetic progression-II, *J. Reine Angew. Math.*, **244** (1970) 108–111.
[29] E. Vegh, A note on the distribution of the primitive roots of a prime, *J. Number Theory*, **3** (1971) 13–18.
[30] E. Vegh, Primitive roots modulo a prime as consecutive terms of an arithmetic progression-III, *J. Reine Angew. Math.*, **256** (1972) 130–137.
[31] S. Wright, Patterns of quadratic residues and nonresidues for infinitely many primes, *J. Number Theory*, **123** (2007) 120–132.
[32] S. Wright, A combinatorial problem related to quadratic non-residue modulo p, To appear Ars Combinatorica.

TCV Proceedings No. 15, 2011, pp. 63–71.

Newforms of Half-Integral Weight on $\Gamma_0(8N)$

M. Manickam

Department of Mathematics, RKM Vivekananda College, Mylapore, Chennai 600 004
Current Address: Kerala School of Mathematics, Kunnamangalam P.O.,
Kozhikode 673 571, India
e-mail: murugumanick@yahoo.com; murugumanick@ksom.res.in

In Honour of Professor Vasudevan's 60th Birthday

1. Introduction

In the year 1973, G. Shimura [9] derived that all the eigenvalues of a Hecke operator T_{n^2} on $S_{k+1/2}(4N)$, the space of cusp forms of weight $k + 1/2$ for the group $\Gamma_0(4N)$, where $k > 1$, $(n, 2N) = 1$, equals a set of eigenvalues of the n-th Hecke operator T_n on $S_{2k}(N')$, the space of cusp forms of weight $2k$, level N'. In 1974, S. Niwa [7] proved that $N' = 2N$. More precisely, if $f(\tau) = \sum_{n\geq 1} a_f(n)e^{2\pi i n\tau}$ is in $S_{k+1/2}(4N)$ and for a square-free integer t, $(-1)^k t \geq 0$ there exists a Hecke-equivariant linear map

$$\mathcal{S}_{t,4N} : S_{k+1/2}(4N) \longrightarrow S_{2k}(2N)$$

$$\mathcal{S}_{t,4N} T_n = T_{n^2}\mathcal{S}_{t,4N} \tag{1}$$

for all $n \geq 1$, $(n, 2N) = 1$, and it is given by

$$f|\mathcal{S}_{t,4N} = \sum_{n\geq 1}\left(\sum_{\substack{d|n\\(d,2N)=1}} d^{k-1}\left(\frac{4t}{d}\right)a_f(|t|n^2/d^2)\right)e^{2\pi i n\tau}.$$

Niwa [8] proved that the trace of T_{n^2} on $S_{k+1/2}(4N)$ equals the trace of T_n on $S_{2k}(2N)$, if N is an odd squarefree positive integer. Since the respective Hecke algebras are semi-simple, there exists a Hecke-equivariant isomorphism between these two spaces. We denote this isomorphism by ψ_N. In the simplest case the space $S_{k+1/2}(4)$ is isomorphic to the space $S_{2k}(2)$. The theory of newforms of integral weight [1] shows that the latter space has the decomposition

$$S_{2k}(2) = (S_{2k}(1) \oplus S_{2k}(1)|B_2) \oplus S_{2k}^{\text{new}}(2), \tag{2}$$

where the space $S_{2k}^{\text{new}}(2)$ is called the space of newforms of weight $2k$, level 2. Now, it is natural to find the subspaces in $S_{k+1/2}(4)$ which correspond to $S_{2k}(1)$, $S_{2k}^{\text{new}}(2)$ respectively. In 1980, W. Kohnen [3] found a canonical subspace in $S_{k+1/2}(4)$, called the plus space, consisting of forms whose n-th Fourier coefficients vanish whenever $(-1)^k n \equiv 2, 3 \pmod 4$. Kohnen introduced a hermitian Hecke operator T_4^+ for the prime 2 on the plus space and proved that the plus space is isomorphic onto the space $S_{2k}(1)$. On doing so he introduced modified Shimura maps, called the Shimura–Kohnen maps indexed by fundamental discriminants, and proved that a certain linear combination of such Shimura–Kohnen maps gives an isomorphism between these spaces. In [4], Kohnen genaralised these results and set up a parallel theory of newforms on the plus space $S_{k+1/2}^+(4N)$, when N is odd and square-free. To do so, he computed the trace of $T_{n^2}^+$ on the plus space $S_{k+1/2}^+(4N)$, where $(n, N) = 1$ and proved that it equals the trace of T_n on the space $S_{2k}(N)$. We call ψ_K the isomorphism between these spaces, which commutes with Hecke operators. In order to identify the subspace of newforms or to extend the theory for the full space of cusp forms of square-free level, namely $S_{k+1/2}(4N)$, in [6] we have used the isomorphism ψ_N as a main tool, and while doing so we realised that the main result for the plus space also follows. The aim of this article is to extend such a theory of newforms on the space $S_{k+1/2}(8N)$, where N is odd and square-free.

2. Notations

2.1 *Basic definitions*

We begin by recalling some basic definitions regarding modular forms of half-integral weight. We refer to Chapter 4 of [2] for more details.

Let $\mathcal{H}$ denote the upper half-plane consisting of complex numbers $z \in \mathbb{C}$ with $\mathrm{Im}(z) > 0$. For complex numbers $0 \neq z, x$, we let $z^x = e^{x \log z}$, $\log z = \log|z| + i \arg z$, $-\pi < \arg z \leq \pi$. Let ξ be a fourth root of unity. Let G denote the four-sheeted covering of $GL_2^+(\mathbb{Q})$ defined as the set of all ordered pairs $(\alpha, \phi(z))$, where $\phi(z)$ is a holomorphic function on $\mathcal{H}$ such that $\phi^2(z) = \xi^2 \frac{cz+d}{\sqrt{\det \alpha}}$ and $\alpha = \left(\begin{smallmatrix} a & b \\ c & d \end{smallmatrix}\right) \in GL_2^+(\mathbb{Q})$. Then G is a group with the multiplicative rule:

$$(\alpha, \phi(z))(\beta, \psi(z)) = (\alpha\beta, \phi(\beta z)\psi(z)).$$

Let $k(\geq 2) \in \mathbb{N}$. For a complex valued function f defined on the upper half-plane $\mathcal{H}$ and an element $(\alpha, \phi(z)) \in G$, define the stroke operator $|_{k+1/2}$ by

$$f|_{k+1/2}(\alpha, \phi(z))(z) = \phi(z)^{-2k-1} f(\alpha z).$$

We omit the subscript $k + 1/2$ and write $|$ instead of $|_{k+1/2}$ for the stroke operator wherever there is no ambiguity. For $\Gamma_0(4)$ and its subgroups, we take the embedding $\Gamma_0(4) \hookrightarrow G$ as the collection $\{(\alpha, j(\alpha, z)), \alpha = \left(\begin{smallmatrix} a & b \\ c & d \end{smallmatrix}\right) \in \Gamma_0(4)\}$, where

$$j(\alpha, z) = \left(\frac{c}{d}\right)\left(\frac{-4}{d}\right)^{-1/2} (cz+d)^{1/2}.$$

Here $\left(\frac{c}{d}\right)$ denotes the generalized quadratic residue symbol and

$$\left(\frac{-4}{d}\right)^{1/2} = \begin{cases} 1 & \text{if } d \equiv 1 \pmod 4, \\ i & \text{if } d \equiv 3 \pmod 4. \end{cases}$$

Let M be a natural number and ψ be an even Dirichlet character modulo $4M$. A holomorphic function $f : \mathcal{H} \longrightarrow \mathbb{C}$ is called a modular form of weight $k + 1/2$ for $\Gamma_0(4M)$ with character ψ, if

$$f|_{k+1/2}(\gamma, j(\gamma, z))(z) = \psi(d) f(z),$$

for all $\gamma = \left(\begin{smallmatrix} a & b \\ c & d \end{smallmatrix}\right) \in \Gamma_0(4M)$ and f is holomorphic and vanishes at all cusps of $\Gamma_0(4M)$. The set of cusp forms defined as above forms a complex vector space denoted by $S_{k+1/2}(4M, \psi)$. When ψ is trivial, we denote this space by $S_{k+1/2}(4M)$. We denote the Fourier expansion of a cusp form f at the cusp $i\infty$ by

$$f(z) = \sum_{n \geq 1} a_f(n) e(nz),$$

where $e(x) = e^{2\pi i x}$, $x \in \mathbb{C}$.

2.2 *Hecke and other operators*

For $(n, 2M) = 1$, let T_{n^2} denote the n-th Hecke operator on $S_{k+1/2}(4M, \psi)$ and for $n \geq 1$, define U_n and B_n by

$$U_n : \sum_{m \geq 1} a(m) e(mz) \mapsto \sum_{m \geq 1} a(nm) e(mz),$$

$$B_n : \sum_{m \geq 1} a(m) e(mz) \mapsto \sum_{m \geq 1} a(m) e(nmz).$$

With respect to the stroke operation, the operators U_n and B_n are given as follows.

$$U_n = n^{\lambda/2-1} \sum_{u \,(\mathrm{mod}\ n)} \left(\begin{pmatrix} 1 & u \\ 0 & n \end{pmatrix}, n^{1/4} \right),$$

$$B_n = n^{-\lambda/2} \left(\begin{pmatrix} n & 0 \\ 0 & 1 \end{pmatrix}, n^{-1/4} \right). \tag{3}$$

The Petersson inner product for cusp forms $f, g \in S_{k+1/2}(4M, \psi)$ is defined by

$$\langle f, g \rangle = \frac{1}{i_{4M}} \int_{\mathcal{F}} f(z) \overline{g(z)} y^{k-3/2} dx dy, \tag{4}$$

where $\mathcal{F}$ is a fundamental domain for the action of $\Gamma_0(4M)$ on $\mathcal{H}$, i_{4M} is the index of $\Gamma_0(4M)$ in $SL_2(\mathbb{Z})$ and $z = x + iy$. We use a similar definition for the Petersson inner product in the case of cusp forms of integral weight.

3. Newforms on $S_{k+1/2}(8N)$

Assume that N is an odd square-free integer. Consider the full space $S_{k+1/2}(8N)$. Let us first introduce certain types of W-operators for the prime $p = 2$.

$$W(8) = \left(\begin{pmatrix} 8a & b \\ 8Nc & 8 \end{pmatrix}, 8^{1/4}e^{\pi i/4}(4Nc\tau + 1)^{1/2}\right), \tag{5}$$

where a, b, c are integers, $b \equiv 1 \pmod 8$ and $8a - Nbc = 1$.

$$W_*(4) = \left(\begin{pmatrix} 4r & s \\ 4Nu & 8 \end{pmatrix}, 4^{1/4}e^{\pi i/4}(-1)^{\left[\frac{k+1}{2}\right]}(Nu\tau + 2)^{1/2}\right), \tag{6}$$

where r, s, u are integers, $s \equiv 1 \pmod 8$ and $8r - Nsu = 1$.

Proposition 3.1. *The action of the operators $W(8)$ and $W_*(4)$ on the space $S_{k+1/2}(8N)$ are independent of the choice of the representatives a, b, c and r, s, u. Moreover, we have*

$$W_*(4) = W(4) \quad \text{on } S_{k+1/2}(4N)$$

and

$$W(8)W_*(4) = (-1)^{\left[\frac{k+1}{2}\right]}\left(\begin{pmatrix} 1 & 0 \\ 0 & 2 \end{pmatrix}, 2^{1/4}\right)$$

on $S_{k+1/2}\left(8N, \left(\frac{8}{\cdot}\right)\right)$. Here

$$W(4) = \left(\begin{pmatrix} 4a & b \\ 4Nc & 4 \end{pmatrix}, 4^{1/4}e^{\pi i/4}(-1)^{\left[\frac{k+1}{2}\right]}(Nc\tau + 1)^{1/2}\right)$$

where a, b, c are integers, $b \equiv 1 \pmod 4$, $4a - Nbc = 1$. It is the W-operator on $S_{k+1/2}(4N)$) for the prime 2.

Proof. For integers a', b', c' with $8a' - Nb'c' = 1$, $b' \equiv 1 \pmod 8$, let

$$W(8)' = \left(\begin{pmatrix} 8a' & b' \\ 8Nc' & 8 \end{pmatrix}, 8^{1/4}e^{\pi i/4}(Nc'\tau + 1)^{1/2}\right).$$

A straightforward computation gives

$$W(8)W(8)'^{-1} = \begin{pmatrix} 8a - Nbc' & a'b - ab' \\ 8N(c - c') & 8a' - Ncb' \end{pmatrix}^*. \tag{7}$$

This proves that the action of $W(8)$ on the space $S_{k+1/2}(8N)$ is independent of the choice of such integers a, b, c. Similar arguments show that $W_*(4)$ on $S_{k+1/2}(8N)$ is also independent of the choice of such integers r, s, u. Now

$$W(8)W_*(4) = \left(4I, (-1)^{\left[\frac{k+1}{2}\right]}\right)\begin{pmatrix} 8ar + Nbu & b + as \\ 8N(rc + u) & 8 + Ncs \end{pmatrix}^*\left(\begin{pmatrix} 1 & 0 \\ 0 & 2 \end{pmatrix}, 2^{1/4}\right), \tag{8}$$

where we have used the following.

$$\left(\frac{8N(rc + u)}{8 + Ncs}\right) = \left(\frac{Ns}{8 + Ncs}\right)\left(\frac{8N(Ncs + 8)r - 1)}{Ncs + 8}\right) = \left(\frac{-8s}{Ncs + 8}\right) = -1, \tag{9}$$

and $\left(\frac{-4}{Ncs+8}\right) = -1$. Hence we have

$$W(8)W_*(4) = (-1)^{\left[\frac{k+1}{2}\right]}\left(\begin{pmatrix}1 & 0\\ 0 & 2\end{pmatrix}, 2^{1/4}\right) \tag{10}$$

on $S_{k+1/2}\left(8N, \left(\frac{8}{\cdot}\right)\right)$. Similar arguments give

$$W(4)W_*(4)^{-1} = \begin{pmatrix}8a - Nbu & br - as\\ 4N(2c-u) & 4r - Ncs\end{pmatrix}^*, \tag{11}$$

and hence, $W(4) = W_*(4)$ on $S_{k+1/2}(4N)$. This complete the proof. □

Proposition 3.2. *The operator $W(8)$ is an isomorphism from $S_{k+1/2}(8N)$ onto $S_{k+1/2}\left(8N, \left(\frac{8}{\cdot}\right)\right)$ and*

$$W(8)T(p^2) = T(p^2)W(8) \quad on \quad S_{k+1/2}(8N)$$

for each prime $(p, 2N) = 1$.

Proof. A simple computation shows that

$$W(8)^2 = \begin{pmatrix}8a^2 + Nbc & b(a+1)\\ 8Nc(a+1) & 8 + Nbc\end{pmatrix}^* = I \quad \text{on } S_{k+1/2}(8N).$$

If $\left(\begin{smallmatrix}\alpha & \beta\\ 8N\gamma & \delta\end{smallmatrix}\right) \in \Gamma_0(8N)$, then

$$\begin{aligned}
&W(8)\begin{pmatrix}\alpha & \beta\\ 8N\gamma & \delta\end{pmatrix}^* W(8)^{-1}\\
&= \left(\begin{pmatrix}0 & -1\\ 8 & 0\end{pmatrix}, 8^{1/4}e^{-\pi i/4}\tau^{1/2}\right)\begin{pmatrix}u & v\\ w & x\end{pmatrix}^* \left(\begin{pmatrix}0 & -1\\ 8 & 0\end{pmatrix}, 8^{1/4}e^{-\pi i/4}\tau^{1/2}\right)^{-1}\\
&= \left(I, \left(\frac{8}{\delta}\right)\right)\begin{pmatrix}x & -w/8\\ -8v & u\end{pmatrix}^*\\
&= \left(I, \left(\frac{8}{\delta}\right)\right) \quad \text{on } S_{k+1/2}(8N),
\end{aligned} \tag{12}$$

where

$$\begin{pmatrix}u & v\\ w & x\end{pmatrix} = \begin{pmatrix}Nc & 1\\ -8a & -b\end{pmatrix}\begin{pmatrix}\alpha & \beta\\ 8N\gamma & \delta\end{pmatrix}\begin{pmatrix}-b & -1\\ 8a & Nc\end{pmatrix}.$$

Using $v \equiv 0 \pmod N$, $w \equiv 0 \pmod 8$, $\left(\begin{smallmatrix}x & -w/8\\ -8v & u\end{smallmatrix}\right) \in \Gamma_0(8N)$. This proves that $W(8)$ maps $S_{k+1/2}(8N)$ into $S_{k+1/2}\left(8N, \left(\frac{8}{\cdot}\right)\right)$. Using explicit representations of $T(p^2)$ we get the commutative property

$$W(8)T(p^2) = T(p^2)W(8) \quad \text{on } S_{k+1/2}(8N),$$

where p is an odd prime relatively prime to N. □

3.1 *Oldforms and newforms*

Let us introduce the following notation.

$$S_{k+1/2;d} = \begin{cases} S^{+}_{k+1/2}(4d) & \text{if } 2 \nmid d, \\ S_{k+1/2}(2d) & \text{otherwise.} \end{cases}$$

The space of oldforms is defined as

$$S^{\text{old}}_{k+1/2}(8N) = \sum_{\substack{rd|4N,\\ d\neq 4N}} S_{k+1/2,d} \big| U(r,d),$$

where

$$U(r,d) = \begin{cases} U(r^2/2)W(8) & \text{if } 4|rd, \\ U(r^2) & \text{if } 4 \nmid d. \end{cases}$$

We define the orthogonal complement of $S^{\text{old}}_{k+1/2}(8N)$ via the Petersson scalar product as the space of newforms. We denote it as $S^{\text{new}}_{k+1/2}(8N)$.

By computing the traces, M. Ueda [10] extended the isomorphism given by Niwa to $S_{k+1/2}(8N)$:

$$\psi_N : S_{k+1/2}(8N) \longrightarrow S_{2k}(4N).$$

We prove a characterisation of the plus space in $S_{k+1/2}(4N)$.

Lemma 3.3. *Let* $f \in S_{k+1/2}(4N)$. *Then*

$$f|U(2)W(8) = (-1)^{\left[\frac{k+1}{2}\right]} 2^{k/2-1/4} f.$$

if and only if $f \in S^{+}_{k+1/2}(4N)$.

Proof. Let $f \in S^{+}_{k+1/2}(4N)$. Then $a_f(n) = 0$ if $n \equiv 2 \pmod 4$. That is, $a_{f|U(2)}(n) = 0$ for all odd n. Hence,

$$f|U(2) = f|U(4)B(2).$$

This implies that

$$\begin{aligned} f|U(2)W(8) &= f|U(4)B(2)W(8) \\ &= (-)^{\left[\frac{k+1}{2}\right]} 2^{-k/2-1/4} f|U(4)W(4) \\ &= (-1)^{\left[\frac{k+1}{2}\right]} 2^{k/2-1/4} f, \end{aligned} \tag{13}$$

where we have used the condition for the plus space: $f|U(4)W(4) = 2^k f$ (see [6, 5]).

Suppose that $f \in S_{k+1/2}(4N)$ satisfies the condition

$$f|U(2)W(8) = (-1)^{\left[\frac{k+1}{2}\right]} 2^{k/2-1/4} f.$$

Then,

$$f|U(2)W(8)W_*(4) = (-1)^{\left[\frac{k+1}{2}\right]} 2^{k/2-1/4} f|W_*(4). \tag{14}$$

$$f|U(2)\left(\begin{pmatrix} 1 & 0 \\ 0 & 2 \end{pmatrix}, 2^{1/4}\right) = 2^{k/2-1/4} f\big|W(4). \tag{15}$$

This implies that $a_{f|U(2)}(n) = 0$ if n is odd, and hence, $a_f(n) = 0$ if $n \equiv 2 \pmod 4$. Thus, $f \in S^+_{k+1/2}(4N)$. This completes the proof.

Using this lemma we prove our main theorem.

Theorem 3.4.

(i) *The space of oldforms has a direct sum decomposition:*

$$S^{\text{old}}_{k+1/2}(8N) = \bigoplus_{rd|4N, d \neq 4N} S^{\text{new}}_{k+1/2;d}\big|U(r,d), \tag{16}$$

where

$$S^{\text{new}}_{k+1/2;d} = \begin{cases} S^{+,\text{new}}_{k+1/2}(4d) & \text{if } 2 \nmid d, \\ S^{\text{new}}_{k+1/2}(2d) & \text{otherwise.} \end{cases}$$

(ii) *The orthogonal complement of the space of old forms $S^{\text{old}}_{k+1/2}(8N)$ in the full space via the Petresson scalar product defines a canonical subspace: $S^{\text{new}}_{k+1/2}(8N)$, and the "strong multiplicity one theorem" holds on $S^{\text{new}}_{k+1/2}(8N)$. Moreover, if N_1, N_2 are distinct odd integers, and if $f_1 \in S^{\text{new}}_{k+1/2}(8N_1)$, $f_2 \in S^{\text{new}}_{k+1/2}(8N_2)$ are Hecke eigenforms under almost all Hecke operators, then the eigenvalues are distinct for infinitely many Hecke operators.*

(iii) *There exists a linear combination of Shimura maps which defines an isomorphism from $S_{k+1/2}(8N)$ onto the space $S_{2k}(4N)$. This maps newforms into newforms and oldforms into oldforms.*

Proof. To derive the result (i) we use induction on N. Consider the case $N = 1$. We prove the required direct sum in the simplest case. If $S^+_{k+1/2}(4)|U(8)W(8)$ contains an element of $S^{\text{old}}_{k+1/2}(4)$, we have f, g_1, g_2 in the plus space $S^+_{k+1/2}(4)$,

$$f|U(8)W(8) = g_1 + g_2|U(4),$$

and hence by applying $W_*(4)$ on both the sides and using Proposition 3.1, and a characterisation of the plus space we get

$$(-1)^{\left[\frac{k+1}{2}\right]} f|U(8)\left(\begin{pmatrix} 1 & 0 \\ 0 & 2 \end{pmatrix}, 2^{1/4}\right) = g_1\big|W(4) + 2^k g_2.$$

Thus, if $n \equiv 2 \pmod 4$, then $a_{f|U(4)}(n) = 0$, and hence $f|U(4) \in S^+_{k+1/2}(4)$. Since, $f \in S^+_{k+1/2}(4)$, $f = 0$. The theory of newforms in $S_{k+1/2}(4)$ (see [6, 5]) shows that $S^+_{k+1/2}(4)$ never contains a non-zero form in $S^{\text{new}}_{k+1/2}(4)$. Finally, if f, $f|U(8)W(8)$ are in $S^{\text{new}}_{k+1/2}(4)$, by applying $W_*(4)$ and using Lemma 3.3 we see that $f \in S^{+}_{k+1/2}(4)$ which is also in $S^{\text{new}}_{k+1/2}(4)$ and hence $f = 0$. Thus, we have the direct sum:

$$\begin{aligned} S^{\text{old}}_{k+1/2}(8) &= S_{k+1/2}(4) + S^{+}_{k+1/2}(4)|U(8)W(8) + S^{\text{new}}_{k+1/2}(4)|U(8)W(8) \\ &= S_{k+1/2}(4) \oplus S^{+}_{k+1/2}(4)|U(8)W(8) \oplus S^{\text{new}}_{k+1/2}(4)|U(8)W(8), \end{aligned} \tag{17}$$

where

$$S_{k+1/2}(4) = S^{+}_{k+1/2}(4) \oplus S^{+}_{k+1/2}(4)|U(4) \oplus S^{\text{new}}_{k+1/2}(4).$$

Next we prove that

$$\psi_N : S^{\text{new}}_{k+1/2}(8N) \longrightarrow S^{\text{new}}_{2k}(4N). \tag{18}$$

Let $f \in S^{\text{new}}_{k+1/2}(8N)$, f be a non-zero Hecke eigenform under all the operators T_{n^2}, $(n, 2N) = 1$. Then we have a square-free integer t, $(-1)^k t > 0$, and

$$F = f|\mathcal{S}_{t,4N}$$

is a non-zero Hecke eigenform in $S_{2k}(4N)$. First we prove that this form is a newform. Using the theory of newforms in $S_{k+1/2}(4N)$, and using Lemma 3.3, we find that

$$\begin{aligned} &\left(S^{+,\text{new}}_{k+1/2}(4N) \oplus S^{+,\text{new}}_{k+1/2}(4N)|U(4) \oplus S^{+,\text{new}}_{k+1/2}(4N)|U(8)W(8)\right)\big|\psi_N \\ &= S^{\text{new}}_{2k}(N) \oplus S^{\text{new}}_{2k}(N)|B(2) \oplus S^{\text{new}}_{2k}(N)|B(4) \end{aligned} \tag{19}$$

Since $F|T_p = \lambda_p F$, (p odd, $p \nmid N$),

$$F|U(p) = \begin{cases} 0 & \text{if } p = 2, \\ \pm p^{k-1} F & \text{if } p \text{ odd}, p|N. \end{cases}$$

Then, either $F \in S^{\text{new}}_{2k}(4N)$ or $F = c_1 G + c_2 G|B(2) + c_3 G|B(4)$, where $G \in S^{\text{new}}_{2k}(N)$ or $G \in S^{\text{new}}_{2k}(2N)$ with $c_3 = 0$. If the latter is true, then $F|\psi_N^{-1} \in S^{\text{old}}_{k+1/2}(4N)$, a contradiction. Hence $F \in S^{\text{new}}_{2k}(4N)$. Now F, $f|\psi_N$ have the same eigenvalues for all T_p, $(p, 2N) = 1$, which proves that $f|\psi_N \in S^{\text{new}}_{2k}(4N)$. Since ψ_N is an Hecke-equivariant isomorphism, we get the strong multiplicity 1 theorem on $S^{\text{new}}_{k+1/2}(8N)$. This completes the proof of (ii). Let us now assume that the direct sum decomposition holds good as in (i) for all square-free integers which are proper divisors of N. With this we prove the result for N. Due to the mapping nature of the isomorphism ψ_N and using the theory of newforms on $S_{k+1/2}(4N)$, (i) follows if we prove that

$$\sum_{rd|4N} S^{\text{new}}_{k+1/2;d}|U(r,d) = \bigoplus_{rd|4N} S^{\text{new}}_{k+1/2;d}|U(r,d), \tag{20}$$

for each r, d with $rd|4N$. Note that

$$S^{\text{old}}_{k+1/2}(8N) = S_{k+1/2}(4N) + \sum_{d|2N} S^{\text{new}}_{k+1/2;d}|U(8)W(8) + \sum_{rd|N} S^{\text{new}}_{k+1/2}(4d)|U(r^2). \tag{21}$$

From Lemma 3.3 we get

$$S_{k+1/2}(4N) + \sum_{d|2N} S^{\text{new}}_{k+1/2,d}|U(8)W(8) = S_{k+1/2}(4M) \bigoplus_{rd|2N} S^{\text{new}}_{k+1/2,d}|U(8)W(8). \tag{22}$$

The induction principle shows that

$$\sum_{rd|N} S^{\text{new}}_{k+1/2}(4d)|U(r^2) = \bigoplus_{rd|N} S^{\text{new}}_{k+1/2}(4d)|U(r). \tag{23}$$

This proves (i). Finally, to prove (iii), we need the following fact whose proof is in [5]. We state it here as a Proposition without proof.

Proposition 3.5. *Let N be any odd square-free integer and let $f \in S^{\text{new}}_{k+1/2}(8N)$ be a non-zero Hecke eigenform for all T_{n^2}, $(n, 2N) = 1$. Then, we have a square-free integer t, $(-1)^k t > 0$ with $N|t$, such that $a_f(t) \neq 0$.*

Invoking this proposition, we find various such square-free integers t_i for a finite set of orthogonal basis of Hecke eigenforms f_i, whose Fourier coefficients $a_{f_i}(|t_i|) \neq 0$. Then as done by Kohnen [4] we find a linear combination of such Shimura maps giving the required isomorphism.

References

[1] A. O. L. Atkin and J. Lehner, Hecke operators on $\gamma_0(m)$, *Math. Ann.*, **69** (1970).

[2] N. I. Koblitz, Introduction to Elliptic Curves and Modular Forms, *Graduates Texts in Mathematics*, **97** Springer-Verlag, New York (1984).

[3] W. Kohnen, Modular forms of half-integral weight on $\Gamma_0(4)$, *Math. Ann.*, **248** (1980) 249–266.

[4] W. Kohnen, Newforms of half-integral weight, *J. Reine Angew Math.*, **333** (1982) 32–72.

[5] M. Manickam, Newforms of half-integral weight and some problems on modular forms, Ph.D thesis, University of Madras (1990).

[6] M. Manickam, B. Ramakrishnan and T. C. Vasudevan, On newforms of half-integral weight, *J. Number Theory*, **32** (1990).

[7] S. Niwa, Modular forms of half-integral weight and integrals of certain theta functions, *Nagoya Math. J.*, **56** (1974) 147–161.

[8] S. Niwa, On Shimra's trace formula, *Nagoya Math. J.*, **66** (1977) 183–202.

[9] G. Shimura, On modular forms of half-integral weight, *Ann. Math.*, **97** (1973) 440–481.

[10] M. Ueda, The decomposition of the spaces of cusp forms of half-integral weight and the trace formula of Hecke operators, *J. Math. Kyoto U.*, **28** (1988) 505–555.

TCV Proceedings No. 15, 2011, pp. 73–87.

On the Zeros of the Epstein Zeta Function

Anirban Mukhopadhyay, Krishnan Rajkumar and Kotyada Srinivas

Institute of Mathematical Sciences, CIT Campus, Tharamani, Chennai 600 113, India
e-mail: anirban@imsc.res.in; srini@imsc.res.in; rkrishnan@imsc.res.in

Abstract. In this article, we count the number of consecutive zeros of the Epstein zeta-function, associated to a certain quadratic form, on the critical line with ordinates lying in $[0, T]$, T sufficiently large and which are separated apart by a given positive number V.

2000 Subject Classification: Primary: 11E45; Secondary: 11M41

1. Introduction

The study of the distribution of zeros of the zeta-functions on the *critical line* is a fundamental problematique in analytic number theory. A related problem is to find a good upper bound for the difference beteween consecutive zeros on the critical line. Such a study natually leads one to ask the following

Question 1. Given T sufficiently large, find $H = H(T)$ such that the interval $[T, T + H]$ contains the ordinate of a zero on the critical line of the corresponding zeta function under consideration?

Question 2. Given T sufficiently large and $V > 0$, how many consecutive zeros of the zeta-function under consideration are there on the critical line with ordinates in $[0, T]$ which are atleast V distance apart?

Question 1 has been studied in great detail from the time of Hardy and Littlewood. We shall briefly mention few known results below. The main object of this note is to study the second question with respect to the Epstein zeta function $\zeta_Q(s)$ (see (2.1)). We shall address this issue in section 2.

In a classical paper [7], Hardy and Littlewood gave an answer to Question 1 for the Riemann zeta function, showing that one can take $H = T^{1/4+\varepsilon}$. This was subsequently improved to $H = T^{1/6}(\log T)^{5+\varepsilon}$ by Moser [17] and to $H = T^{1/6+\varepsilon}$ by Balasubramanian [1], and the latest result is by Karatsuba [15] who showed that $H = T^{5/32}(\log T)^2$ holds. It must be mentioned here that Ivić (page 261, [9]) improved Karatsuba's exponent $5/32 = 0.15625$ to $0.1559458\ldots$. Following the method of Hardy and Littlewood, Potter and Titchmarsh obtained analogous result for the Epstein zeta-function, which we shall discuss in section 2. For the zeta function, $\zeta_K(s, C)$ of an ideal class C in a quadratic number field K, Bruce Bernt [3] showed that $H = T^{1/2+\varepsilon}$ holds. This was improved to $H = T^{1/2}\log T$ by Sankaranarayanan [22]. In [11], Jutila developed a method of transforming certain exponential sums into another sum which is much easier to handle (see section 4) and thereby proved that $H = T^{1/3+\varepsilon}$ holds for the L-function associated to a cusp form (1.3) for the full modular group.

A well-known theorem of Selberg [23] states that the function $\zeta\left(\frac{1}{2}+it\right)$ has at least $\gg T \log T$ zeros in the interval $[0, T]$, and this is best possible upto a value of the implied constant. In fact, Heath-Brown [8] showed that the same estimate holds even if we count only simple zeros. Hence the average gap between critical zeros of $\zeta(s)$ is $(\log T)^{-1}$. It is in this context that the Question 2 becomes relevant!

Let $R_1 := R_1(V)$ be the number of gaps between consecutive zeros of $\zeta\left(\frac{1}{2}+it\right)$, with ordinate between 0 and T, which are larger than V. Then trivially $R_1 \ll TV^{-1}$. Karatsuba [16] showed that $R_1 \ll TV^{-3/2}$ for $V = T^{\delta}$ for any $\delta > 0$. Ivić and Jutila [10] made substantial improvement in the case of $\zeta(s)$ as well as the L-function associated to a certain cusp form. We state their results below as Theorem A and B:

Theorem A. *Let R_1 be as defined above. Then uniformly*

$$R_1 \ll TV^{-2} \log T \tag{1.1}$$

and

$$R_1 \ll TV^{-3} \log^5 T. \tag{1.2}$$

For $V \ll \log T$ these results are trivial and also $R_1 = 0$ for $V = T^{1/6}$ which follows from known results mentioned earlier on the gap between critical zeros for $\zeta(s)$. Thus, the result is non-trivial in the range $\log T \ll V \leq T^{1/6-\epsilon}$. In this article we shall be always considering V in the range $T^{\epsilon} \ll V \leq T^{1/6-\epsilon}$.

Let $a(n)$'s denote the Fourier coefficients of a cusp form f of weight κ for the full modular group. The L-series associated with f is given by the Dirichlet series

$$\varphi(s) = \sum_{n=1}^{\infty} \frac{a(n)}{n^s}, \qquad Re(s) > (\kappa+1)/2. \tag{1.3}$$

Let $R_2 := R_2(V)$ denote the number of gaps of length at least V between consecutive zeros of $\varphi(\kappa/2+it)$ in the interval $[0, T]$.

Theorem B. *Suppose that the $a(n)$'s defined above are real and let $V \gg \log^5 T$. Then*

$$R_2 \ll TV^{-2} \log T \tag{1.4}$$

and

$$R_2 \ll T^2 V^{-6} \log^6 T. \tag{1.5}$$

In this note, we obtain an analogue of the results (1.1) and (1.4) for the Epstein zeta-function, $\zeta_Q(s)$. We state the main result in the next section after a brief survey of the known results on the zeros of $\zeta_Q(s)$. In section 3 and 4 we shall state some Lemmas, section 5 will deal with the proof of the main result.

2. Main result

Let $Q(x, y) = ax^2 + bxy + cy^2$ be a binary positive-definite integral quadratic form. Let $r_Q(n)$ be the number of solutions of $Q(x, y) = n$. The Epstein zeta-function associated to $Q(x, y)$ is defined as

$$\zeta_Q(s) = \sum_{(x,y)\neq(0,0)} \frac{1}{Q(x, y)^s} = \sum_{n=1}^{\infty} \frac{r_Q(n)}{n^s}, \qquad \sigma > 1. \tag{2.1}$$

It is well known that $\zeta_Q(s)$ can be analytically continued to the entire complex plane except for $s = 1$ where it has a simple pole with residue $2\pi/\sqrt{\Delta}$ where $\Delta = 4ac - b^2 > 0$. It also has the functional equation

$$\left(\frac{\sqrt{\Delta}}{2\pi}\right)^s \Gamma(s)\zeta_Q(s) = \left(\frac{\sqrt{\Delta}}{2\pi}\right)^{1-s} \Gamma(1-s)\zeta_Q(1-s). \tag{2.2}$$

We suppose throughout the paper that Δ is not a square, so that $\sqrt{\Delta}$ is irrational; other cases like $\Delta = 4$ related to the form $Q(x, y) = x^2 + y^2$ are either easier or well-known.

Let $h(-\Delta)$ be the number of inequivalent forms with discriminant $-\Delta$. It is well known that there is a one-one correspondence between the equivalence classes of quadratic forms of discriminant $-\Delta$ and the ideal classes of $K = \mathbb{Q}(\sqrt{-\Delta})$. In fact, the Dedekind zeta-function $\zeta_K(s)$ for the number field K can be written as

$$\zeta_K(s) = \frac{1}{l} \sum_r \zeta_{Q_r}(s),$$

where Q_r runs over the distinct equivalence classes and l is the number of units in K.

The Dedekind zeta-function has an Euler product and it belongs to that class of Dirichlet series for which Riemann's hypothesis can be reasonably expected. In contrast, if the class number $h(-\Delta)$ of the quadratic forms with discriminant $-\Delta$ exceeds one, then $\zeta_Q(s)$ has no Euler product and the Riemann hypothesis fails to hold for it. Indeed, when a, b, c are integers, $-\Delta$ is a fundamental discriminant and the class number $h(-\Delta) > 1$, Davenport and Heilbronn [6] had shown that $\zeta_Q(s)$ has infinitely many zeros in the half plane $\sigma > 1$ aribitrarily close to the line $\sigma = 1$. S. M. Voronin [27] proved that if $h(-\Delta) > 1$, then the number of zeros of $\zeta_Q(s)$ in the rectangle $\sigma_1 \le \sigma \le \sigma_2$, $|t| \le T$, with $1/2 < \sigma_1 < \sigma_2 \le 1$ and T sufficiently large, exceeds $c(\sigma_1, \sigma_2)T$, where $c(\sigma_1, \sigma_2) > 0$. Chowla and Selberg [5] showed the existence of a real zero s with $\frac{1}{2} < s < 1$ when $a = 1, b = 0$ and c is large enough. Bateman and Grosswald [2] improved it by showing that $\zeta_Q(s)$ has a real zero between $\frac{1}{2}$ and 1 if $k = (\sqrt{\Delta})/2a > 7.0556$.

Stark [25] gave a striking description about the distribution of zeros of $\zeta_Q(s)$ in a bounded region. More precisely, he proved that there exists a number K such that if $k > K$, then all the zeros of $\zeta_Q(s)$ in the region $-1 < \sigma < 2$, $-2k \le t \le 2k$ are simple zeros and with the exception of two real zeros between 0 and 1, all are on the line $\sigma = \frac{1}{2}$. Moreover, in the same paper he proved the following zero-density result: Let $N(T, Q)$ denote the number of zeros of $\zeta_Q(s)$ in the region $-1 < \sigma < 2$, $0 \le t \le T$. If $k > K$ and $0 < T \le 2k$, then

$$N(T, Q) = (T/\pi)\log(kT/\pi e) + O(h(T+3)),$$

where $h(x) = (\log x)^{1/3}(\log\log x)^{1/6}$ and the constant in "O" does not depend on k.

Though $\zeta_Q(s)$, in general, has thus infinitely many zeros *off* the critical line $\sigma = 1/2$, it has nevertheless infinitely many zeros *on* the critical line, and in fact it was proved by Potter and Titchmarsh [20] that there is a zero $1/2 + i\gamma$ of ζ_Q with $\gamma \in [T, T + T^{1/2+\varepsilon}]$, for any fixed $\varepsilon > 0$ and $T \ge T(Q, \varepsilon)$. Sankaranarayanan [22] sharpened this by showing the same for intervals of the type $[T, T + cT^{1/2}\log T]$. For a long time improving the exponent of T below $1/2$ was considered to be a challenging problem. First result in this direction is by Jutila and Srinivas [14], where they prove that for any positive definite binary integral quadratic form Q and for any fixed $\varepsilon > 0$ and $T \ge T(\varepsilon, Q)$, there is a zero $1/2 + i\gamma$ of the corresponding Epstein zeta-function ζ_Q with

$$|\gamma - T| \le T^{5/11+\varepsilon}. \tag{2.3}$$

Thus the best answer to Question 1 for ζ_Q is $H = T^{5/11+\varepsilon}$, at the time of writing this manuscript. However, in analogy with the L-functions associated to a cusp forms (mentioned in section 1), the next big challenge seems to be to reduce the exponent of T from $5/11$ to $1/3$!

In this article, we prove the following

Theorem. *Suppose V satisfies*

$$T^{\epsilon} \ll V \ll T^{1/2-\epsilon}$$

and $R := R(V)$ denote the number of gaps of length at least V between consecutive zeros of $\zeta_Q\left(\frac{1}{2} + it\right)$ in the interval $[0, T]$, *then*

$$R \ll T^{1+\epsilon}V^{-2} \tag{2.4}$$

for sufficiently large T, where the constant in $\ll$ may depend only on Δ and ϵ.

Remark. The method of the proof closely follows the paper by Ivić and Jutila [10]. Using their approach to find an analogue of (1.2) and (1.5) for $\zeta_Q(s)$ does not seem to yield anything better than (2.4). Note that an analogue of (1.2) or (1.5) would immediately improve the exponent of T to $1/3$ in (2.3).

3. Some preliminary lemmas

Ramachandra showed (see [21], Chapter II) that the first power mean of a generalized Dirichlet series satisfying certain conditions can not be too small. The following Lemma (Theorem 3 of [1]) is a particular case of this general theorem, which is quite useful in obtaining lower bounds of this type, even in short-intervals.

Lemma 1. *Let $B(s) = \sum_{n=1}^{\infty} b_n n^{-s}$ be any Dirichlet series satisfying the following conditions:*

(i) *not all* b_n*'s are zero;*
(ii) *the function can be continued analytically in* $\sigma \geq a$, $|t| \geq t_0$, *and in this region* $B(s) = O((|t| + 10)^A)$.

Then for every $\epsilon > 0$, *we have*

$$\int_T^{T+H} |B(\sigma + it)| dt \gg H$$

for all $H \geq (\log T)^\epsilon$, $T \geq T_0(\epsilon)$, *and* $\sigma > a$.

A suitable Dirichlet polynomial approximation is required to replace the divergent series $\zeta_Q(1/2 + it)$ in the course of the proof. We state the following Lemma which is a direct adaptation of Lemma 3 in [11].

Lemma 2. *Let* $t \geq 2$ *and* $t^2 \ll X \ll t^A$, *where* A *is an arbitrarily large positive constant. Then we have*

$$\begin{aligned}
&\zeta_Q(1/2 + it) \\
&\quad = \sum_{n \leq X} r_Q(n) n^{-1/2-it} + (\log 2)^{-1} \sum_{X < n \leq 2X} r_Q(n) \log(2X/n) n^{-1/2-it} \\
&\qquad + (\log 2)^{-1} 2\pi \Delta^{-1/2} (1/2 - it)^{-2} ((2X)^{1/2-it} - X^{1/2-it}) + O(tX^{-1/2}).
\end{aligned} \tag{3.1}$$

The following result is Lemma 4.3 of Titchmarsh [26].

Lemma 3. *Let* $F(x)$ *and* $G(x)$ *be real functions,* $G(x)/F(x)$ *monotonic, and* $F''(x)/G(x) \geq m > 0$, *or* $\leq -m < 0$. *Then*

$$\left| \int_a^b G(x) e^{iF(x)} dx \right| \leq \frac{4}{m} \tag{3.2}$$

4. Summation and transformation formulae

Let

$$S = \sum_n \eta(n) r_Q(n) n^{-1/2-it}$$

and $e_k(x) = \exp(2\pi x/k)$. Introducing an additive character (mod k), we may write it formally as

$$S = \sum_n \eta(n) r_Q(n) n^{-1/2-it} \mathrm{e}_k(nh) \cdot \mathrm{e}_k(-nh); \tag{4.1}$$

as in [12], the purpose of the extra exponential factor is to damp the oscillations of the original exponential sum before an application of the Voronoi summation. Thus S is of the general form

$$\sum_n r_Q(n) f(n) \mathrm{e}_k(hn),$$

and a Voronoi summation formula for such sums was given in [13], Eq. (28). To state it, we need some notation. The Gauss sums related to the form Q and additive characters are

$$G_Q(k,h) = \sum_{x,y \pmod{k}} e_k(hQ(x,y)),$$

and it holds (see [24], Lemma 1)

$$|G_Q(k,h)| \le (\Delta,k)k. \tag{4.2}$$

Further, the summation formula involves an integral positive definite quadratic form $Q^*(x,y)$ depending on Q and k, and the discriminant of Q^* is at most Δ in absolute value. Also, there occurs an arithmetic function (corresponding to $\tilde{r}_{Q^*}(n)$ in [13]) of the form

$$\rho(n) = \rho(n; Q, h/k) = \sum_{Q^*(x,y)=n} \alpha(x,y), \tag{4.3}$$

where $|\alpha(x,y)| \le (\Delta,k) \le \Delta$ and $\alpha(x,y)$ depends only on the classes of x and y (mod Δ) for given Q and h/k. In this notation, a slightly simplified version of the summation formula (28) in [13] can be stated as follows.

Let $\chi(s)$ be as in the functional equation $\zeta(s) = \chi(s)\zeta(1-s)$, thus $\chi(s) = 2^s\pi^{s-1}\sin\left(\frac{1}{2}\pi s\right)\Gamma(1-s)$. The following Lemma is due to Jutila and Srinivas (Lemma 3.2 [14])

Lemma 4. *Let t be a large positive number, $r = h/k$ a positive rational number with $(h,k) = 1$, and suppose that the positive numbers M_1 and M_2 satisfy*

$$M_1 < \frac{t}{2\pi r} < M_2$$

$$M_j = \frac{t}{2\pi r} + (-1)^j m_j, \quad m_1 \asymp m_2,$$

$$1 \le k \ll {M_1}^{1/2-\delta_1},$$

$$t^{\delta_2}\max(t^{1/2}r^{-1}, hk) \ll m_1 \ll {M_1}^{1-\delta_3}$$

for some small positive constants δ_j. Further, let

$$U \gg r^{-1}t^{1/2+\delta_4}$$

and let J be a fixed positive integer exceeding a certain bound which depends on δ_4. Write

$$M'_j = M_j + (-1)^{j-1}JU = \frac{t}{2\pi r} + (-1)^j m'_j,$$

supposing that $m_j \asymp m'_j$, and let

$$n_j = \Delta_0 h^2 {m_j}^2 {M_j}^{-1}, \quad n'_j = \Delta_0 h^2 (m'_j)^2 (M'_j)^{-1}. \tag{4.4}$$

Then for a certain weight function $\eta \in C^{J-1}(\mathbb{R})$ with support $[M_1, M_2]$ and satisfying $\eta(x) = 1$ for $x \in [{M_1}', {M_2}']$, we have

$$\sum_{n=1}^{\infty} \eta(n) r_Q(n) n^{-1/2-it}$$
$$= \Bigg\{ 2\pi \Delta^{-1/2} k^{-2} G_Q(k, -h) r^{-1/2}$$
$$+ \pi^{1/4} i (2hkt\Delta_0)^{-1/4} (\Delta_0/\Delta)^{1/2} \sum_{j=1}^{2} (-1)^j \sum_{n<n_j} w_j(n) \rho(n)$$
$$\times \exp\left(2\pi i n \left(\frac{\bar{h}\bar{\Delta}_0}{k} - \frac{1}{2hk\Delta_0}\right)\right) n^{-1/4} \left(1 + \frac{\pi n}{2hk\Delta_0 t}\right)^{-1/4}$$
$$\times \exp\left(i(-1)^{j-1}\left(2t\phi\left(\frac{\pi n}{2hkt\Delta_0}\right) + \frac{\pi}{4}\right)\right)\Bigg\} r^{it} \chi\left(\frac{1}{2} + it\right)$$
$$+ O(h^2 k^{-1} m_1{}^{1/2} t^{-3/2} U \log t), \tag{4.5}$$

where

$$\phi(x) = \operatorname{arcsinh}(x^{1/2}) + (x + x^2)^{1/2}, \tag{4.6}$$

$w_j(n) = 1$ *for* $n < n'_j$, $w_j(n) \ll 1$ *for* $n \leq n_j$, $w_j(y)$ *and* $w'_j(y)$ *are piecewise continuous in the interval* (n'_j, n_j) *with at most* $J - 1$ *jumps, and*

$$w'_j(y) \ll (n_j - n'_j)^{-1} \quad \text{for } n'_j < y < n_j$$

whenever $w'_j(y)$ *exists.*

5. Proof of the theorem

We define the functions $f(s)$, $\gamma(s)$ and $W(t)$ as

$$f(s) = e^{\frac{1}{2}\pi i\left(\frac{1}{2}-s\right)} \left(\frac{\sqrt{\Delta}}{2\pi}\right)^s \Gamma(s) \zeta_Q(s) = \gamma(s) \zeta_Q(s)$$

and

$$W(t) = f(1/2 + it);$$

the latter is an analogue of Hardy's function $Z(t)$ in the theory of Riemann's zeta-function. The functional equation for $\zeta_Q(s)$ implies that $W(t)$ is real for real t. Thus the zeros of $\zeta_Q(s)$ on the critical line correspond to the real zeros of $W(t)$. To prove the result we can obviously restrict ourselves to the zeros lying in $[T/2, T]$. From now onwards c_i's will always denote absolute constants. Suppose now that τ and $\tau + U$ are the ordinates of two consecutive zeros of $\zeta_Q(1/2 + it)$ and hence two consecutive zeros of $W(t)$ such that

$$[\tau, \tau + U] \subset [T/2, T], \quad \text{and} \quad U \geq V, \tag{5.1}$$

where $T^{\epsilon} \leq V \leq T^{5/11}$. Further, let $L = 8(\log T)^{1/2}$ and $G = VL^{-1}$. Then, for $t \in [\tau + V/4, \tau + 3V/4]$, $W(u)$ has no zero in the interval $[t - V/4, t + V/4]$ and so $W(u)$ does not change sign in this interval. Let us define

$$I_1(t) = \int_{t-V/4}^{t+V/4} |W(u)|e^{-(t-u)^2/G^2} du$$

and

$$I_2(t) = \int_{t-V/4}^{t+V/4} W(u)e^{-(t-u)^2/G^2} du$$

Then for $t \in [\tau + V/4, \tau + 3V/4]$, $I_1(t) = |I_2(t)|$. Let $\mathcal{S}$ be the set of $t \in [T/2 + V/4, T - V/4]$ such that $I_1(t) = |I_2(t)|$. Since the number of pairs of consecutive zeros of $\zeta_Q(1/2 + it)$, τ and $\tau + U$, such that (5.1) holds is R, the above discussion implies that the Lebesgue measure of $\mathcal{S}$, $m(\mathcal{S}) \gg RV$. Thus

$$RV \ll m(\mathcal{S}) = \int_{t\in\mathcal{S}} dt = \int_{t\in\mathcal{S}} \frac{|I_2(t)|^2}{I_1(t)^2} dt. \tag{5.2}$$

Therefore, to obtain an upper bound on R, we have to find a lower bound for $I_1(t)$ and an upper bound for $I_2(t)$.

The lower bound

$$I_1(t) \gg G. \tag{5.3}$$

follows from Lemma 1.

From now onwards we shall dwell on the upper bound estimation of $I_2(t)$. First we define some notations; let $P(u) = u\sqrt{\Delta}/2\pi$, for J sufficiently large, we choose a smoothing function $\eta \in C^{J-1}(\mathbb{R})$ satisfying the following

$$\eta(x) = \begin{cases} 1 & \text{if } x \in [P(T) - Y, P(T) + Y] \\ 0 & \text{if } x > P(T) + 2Y \quad \text{or } x < P(T) - 2Y \end{cases} \tag{5.4}$$

The parameter Y is defined by

$$VY = T^{1+\epsilon}. \tag{5.5}$$

We shall use e^x and $\exp(x)$ to denote the same function. Now the integral $I_2(t)$ is written as

$$I_2(t) = \int_{-V/4}^{V/4} \gamma(1/2 + i(t+u))\zeta_Q(1/2 + i(t+u))e^{-(u/G)^2} du. \tag{5.6}$$

We substitute the expression for $\zeta_Q(1/2+i(t+u))$ given by Lemma 2 with $X = t^3$ in (5.6) to obtain

$$I_2(t) = \sum_{\substack{|n-P(t)|>2Y \\ n\le t^3}} r_Q(n)n^{-1/2} \int_{-V/4}^{V/4} \gamma(1/2 + i(t+u))n^{-i(t+u)} e^{-(u/G)^2} du$$

$$+ \int_{-V/4}^{V/4} \gamma(1/2 + i(t+u)) \left(\sum_{|n-P(t)|\le 2Y} r_Q(n)n^{-1/2-i(t+u)} \right) e^{-(u/G)^2} du$$

$$+ (\log 2)^{-1} \sum_{t^3 < n \le 2t^3} r_Q(n) \log(2t^3/n) n^{-1/2}$$

$$\times \int_{-V/4}^{V/4} \gamma(1/2 + i(t+u)) n^{-i(t+u)} e^{-(u/G)^2} du + O(Gt^{-1/2})$$

$$= S_1 + S_2 + S_3 + o(G). \tag{5.7}$$

We expand the sum within brackets in S_2 and use (5.4) to get

$$S_2 = \int_{-V/4}^{V/4} \gamma(1/2 + i(t+u)) \left(\sum_{|n-P(t)| \le 2Y} \eta(n) r_Q(n) n^{-1/2-i(t+u)} \right.$$

$$\left. + \sum_{|n-P(t)| \le 2Y} (1 - \eta(n)) r_Q(n) n^{-1/2-i(t+u)} \right) e^{-(u/G)^2} du$$

$$= \int_{-V/4}^{V/4} \gamma(1/2 + i(t+u)) \sum_{n=1}^{\infty} \eta(n) r_Q(n) n^{-1/2-i(t+u)} e^{-(u/G)^2} du$$

$$+ \sum_{Y < |n-P(t)| \le 2Y} (1 - \eta(n)) r_Q(n) n^{-1/2}$$

$$\times \int_{-V/4}^{V/4} \gamma(1/2 + i(t+u)) n^{-i(t+u)} e^{-(u/G)^2} du$$

$$= S_4 + S_5. \tag{5.8}$$

We now estimate S_1, S_3 and S_5. Note that in all these cases $|n - P(t)| > Y$ and $n \ll t^3$. First we consider $n > P(t) + Y$. We view the integral

$$\int_{-V/4}^{V/4} \gamma(1/2 + i(t+u)) n^{-i(t+u)} e^{-(u/G)^2} du$$

appearing in S_1, S_3, S_5, as a complex integral over the rectangular contour with vertices $\pm V/4$, $\pm V/4 - iG$. We then use the well known Stirling's formula for the Γ-function which states that in any fixed vertical strip $-\infty < \alpha \le \sigma \le \beta < \infty$,

$$\Gamma(\sigma + it) = (2\pi)^{1/2} t^{\sigma+it-1/2} e^{-\frac{\pi}{2}t + \frac{\pi}{2}i(\sigma-1/2)-it}(1 + O(1/t)) \quad \text{as } t \to \infty,$$

in order to estimate $\gamma(1/2 + i(t+u))$ along this contour.

Therefore, on the vertical sides, where $u = \pm V/4 - iw$ with $w \in [0, G]$, we estimate the factors in the integrand as follows

$$\gamma(1/2 + i(t+u)) n^{-i(t+u)}$$

$$= \Delta^{1/4} \left(\frac{2\pi n}{\sqrt{\Delta}(t + \frac{V}{4})} \right)^{-w}$$

$$\times \exp\left\{ i\left(\left(t+\frac{V}{4}\right)\log\left(\frac{\sqrt{\Delta}}{2\pi n}\left(t+\frac{V}{4}\right)\right)-\left(t+\frac{V}{4}\right)\right)\right\}(1+O(1/T))$$

$$\ll \left(\frac{n}{P(t)+\frac{\sqrt{\Delta}}{8\pi}V}\right)^{-w}$$

which is bounded. As for the other factor, we have

$$e^{-(u/G)^2} \ll e^{-(L/4)^2} = e^{-4\log T}.$$

Hence the value of the integral on the vertical sides becomes $O(G/T^4)$.

On the horizontal side in the lower half plane, let $u = v - iG$ with $v \in \left[-\frac{V}{4}, \frac{V}{4}\right]$. Here $e^{-(u/G)^2}$ is bounded and the other factor is estimated by

$$\begin{aligned}\gamma(1/2+i(t+u))n^{-i(t+u)} &\ll \exp\left(-G\log\left(\frac{n}{P(t)+\frac{\sqrt{\Delta}}{8\pi}V}\right)\right)\\ &\ll \exp\left(-\frac{GY}{P(T)}\right) = \exp\left(-c\frac{T^{\epsilon}}{L}\right) \ll \exp(-\log(T^4L))\end{aligned}$$

where the constant $c = \sqrt{\Delta}/(2\pi)$. Hence the value of the integral on the horizontal side is also $O(G/T^4)$.

For $n < P(t) - Y$ we take the rectangular contour with vertices $\pm V/4$, $\pm V/4 + iG$ in the upper half plane and argue in a similar way to estimate the value of the integral to be $O(G/T^4)$.

Finally we use the fact that $r_Q(n) \ll n^{\epsilon_1}$ for any $\epsilon_1 > 0$ (see the discussion preceding (5.21)) and estimate S_1, S_3, and S_5 as follows

$$S_j \ll GT^{-4}\sum_{n\le 2t^3} r_Q(n)n^{-1/2} \ll GT^{-4+3/2+\epsilon_1} = o(G) \quad j = 1, 3, 5. \tag{5.9}$$

From equations (5.7), (5.8) and (5.9), after a change of variable, we conclude that

$$I_2(t) \ll \int_{t-V/4}^{t+V/4}\left|\sum_{n=1}^{\infty}\eta(n)r_Q(n)n^{-1/2-iu}\right|e^{-(t-u)^2/G^2}du + o(G). \tag{5.10}$$

Now we need to use the transformation formula (4.5) of Lemma 4 to transform the sum in (5.10). For that we first choose a rational approximation r to $1/\sqrt{\Delta}$ as follows. Since $\sqrt{\Delta}$ is a quadratic irrational, we can choose $r = h/k$ with $(h,k) = 1$, such that $h^{-2} \ll |\sqrt{\Delta} - k/h| \ll h^{-2}$ and $h \asymp \sqrt{T/Y}$. We also observe that $k \asymp h$ as $h/k = r \asymp 1/\sqrt{\Delta}$, which is a constant. Let $M_j = P(u) + (-1)^j 2Y$, $j = 1, 2$. Then

$$m_j = (-1)^j\frac{u}{2\pi}\left(\sqrt{\Delta}-\frac{1}{r}\right) + 2Y \asymp \max(Th^{-2}, Y) \ll Y.$$

Also we choose

$$M'_j = P(u) + (-1)^j Y, \quad m'_j = (-1)^j \frac{u}{2\pi}\left(\sqrt{\Delta} - \frac{1}{r}\right) + Y \ll Y.$$

We observe that $M_j, M'_j \asymp T$. It is easy to check that all conditions of the Lemma 4 are satisfied when $T^{1/2+\epsilon} \ll Y \ll T^{1-\epsilon}$ or in other words when $T^{\epsilon} \ll V \ll T^{1/2-\epsilon}$. Hence by Lemma 4, there exists a smoothing function η satisfying (5.4) for which (4.5) holds. Finally we also get the following estimate for $n_j(u)$,

$$n_j(u) \ll \frac{h^2 m_j^2}{M_j} \ll Y. \tag{5.11}$$

Now we replace the sum in (5.10) by using (4.5). Then the leading term in (4.5) is $\ll k^{-2}|G_Q(k,-h)| \ll k^{-1}$ which after integration in (5.10) becomes $o(G)$. Similarly, the error term in (4.5) is $\ll h^2 k^{-1} Y^{1/2} T^{-3/2} U \log T \ll T^{-1/2+\epsilon}$ which, after integration in (5.10), also becomes $o(G)$.

Thus we obtain the following upper bound on $I_2(t)$,

$$I_2(t) \ll \int_{t-V/4}^{t+V/4} |\Sigma(u)| e^{-(t-u)^2/G^2} du + o(G), \tag{5.12}$$

where $\Sigma(u)$ is as follows

$$\Sigma(u) = (hku)^{-1/4} r^{iu} \chi(1/2+iu) \sum_{j=1}^{2} (-1)^j \sum_{n<n_j(u)} \omega_j(n)\rho(n) n^{-1/4}$$
$$\times \exp(2\pi i C_1 n)(1+C_2 n/u)^{-1/4} \exp(i(-1)^{j-1}(2u\phi(C_2 n/u)+\pi/4)), \tag{5.13}$$

where $C_1 = \bar{h}\bar{\Delta}_0/k - 1/(2hk\Delta_0)$, and $C_2 = \pi/(2hk\Delta_0)$.

We now claim that for $t \in \mathcal{S}$ we can ignore the error term in (5.12). This is because for these t we have (by (5.3)) $G \ll I_1(t) = |I_2(t)|$. Hence we actually have

$$I_2(t) \ll \int_{t-V/4}^{t+V/4} |\Sigma(u)| e^{-(t-u)^2/G^2} du. \tag{5.14}$$

Using the lower bound (5.3) of $I_1(t)$ and the upper bound (5.14) of $I_2(t)$ in (5.2) we get

$$RV \ll G^{-2} \int_{t\in\mathcal{S}} \left| \int_{t-V/4}^{t+V/4} |\Sigma(u)| e^{-(t-u)^2/G^2} du \right|^2 dt. \tag{5.15}$$

Applying the Cauchy-Schwarz inequality to the inner integral, (5.15) becomes

$$RV \ll G^{-2} \int_{t\in\mathcal{S}} \left(\int (|\Sigma(u)| e^{-\frac{1}{2}(t-u)^2/G^2})^2 du \int (e^{-\frac{1}{2}(t-u)^2/G^2})^2 du \right) dt, \tag{5.16}$$

where the inner integrals are over the range $[t - V/4, t + V/4]$. Note that

$$\int_{t-V/4}^{t+V/4} e^{-(t-u)^2/G^2} du = \int_{-V/4}^{V/4} e^{-x^2/G^2} dx \leq \sqrt{\pi} G. \tag{5.17}$$

Hence (5.16) and (5.17) yields

$$RV \ll G^{-1} \int_{t \in \mathcal{S}} \int_{t-V/4}^{t+V/4} |\Sigma(u)|^2 e^{-(t-u)^2/G^2} du.$$

By replacing $\mathcal{S}$ by the bigger set $[T/2 + V/4, T - V/4]$ and interchanging the order of integration we get

$$RV \ll G^{-1} \int_{T/2}^{T} |\Sigma(u)|^2 \int_{u-V/4}^{u+V/4} e^{-(t-u)^2/G^2} dt\, du \ll \int_{T/2}^{T} |\Sigma(u)|^2 du. \tag{5.18}$$

Using (5.13) in (5.18), we have the following bound

$$R \ll V^{-1}(hkT)^{-1/2} \sum_{j=1}^{2} \int_{T/2}^{T} \Bigg| \sum_{n < n_j(u)} \omega_j(n)\rho(n) n^{-1/4} \exp(2\pi i C_1 n)$$

$$\times (1 + C_2 n/u)^{-1/4} \exp(i(-1)^{j-1} 2u\phi(C_2 n/u)) \Bigg|^2 du. \tag{5.19}$$

Suppose that $n_j(T_0) = \max\{n_j(u) : u \in [T/2, T]\}$. By expanding the sum within the modulus in (5.19) we get

$$\begin{aligned} R &\ll V^{-1}(hk)^{-1/2} T^{1/2} \sum_{j=1}^{2} \sum_{n < n_j(T_0)} |\omega_j(n)\rho(n)|^2 n^{-1/2} \\ &\quad + V^{-1}(hkT)^{-1/2} \sum_{j=1}^{2} \sum_{\substack{m,n < n_j(T_0) \\ m \neq n}} |\omega_j(m)\rho(m)\omega_j(n)\rho(n)|(mn)^{-1/4} \\ &\quad \times \int ((1 + C_2 m/u)^{-1/4}(1 + C_2 n/u)^{-1/4} \\ &\quad \times \exp(i(-1)^{j-1} 2u(\phi(C_2 m/u) - \phi(C_2 n/u)))du \\ &= \Sigma_1 + \Sigma_2. \end{aligned} \tag{5.20}$$

In Σ_2, the integral will be over an appropriate subinterval of $[T/2, T]$ depending on m and n.

For estimating Σ_1 we will need an estimate on the mean-square of the coefficients $r_{Q^*}(n)$. Since $r_{Q^*}(n)$ is a complex linear combination of the Dirichlet coefficients of the Dedekind zeta-function $\zeta_K(s)$, we have, for any $\epsilon > 0$, $r_{Q^*}(n) \ll n^{\epsilon}$ for sufficiently large n depending on ϵ. Therefore the following estimate trivially holds.

$$\sum_{n\le x} r_{Q^*}^2(n) = O(x^{1+\epsilon}), \tag{5.21}$$

for any $\epsilon > 0$. In fact, for certain quadratic forms much better results are known (see [18], [19]). Since $\rho(n) \ll r_{Q^*}(n)$ by (4.3), (5.21) implies that the function $m(x)$ defined by

$$m(x) = \sum_{n\le x} |w_j(n)\rho(n)|^2 \ll x^{1+\epsilon} \tag{5.22}$$

for $x \le n_j(T_0)$. Hence, by partial summation we get that the inner sum over n in Σ_1 is

$$\begin{aligned}\sum_{n<n_j} |w_j(n)\rho(n)|^2 n^{-1/2} &= \int_0^{n_j} x^{-1/2} dm(x)\\ &= m(n_j)n_j^{-1/2} + \frac{1}{2}\int_0^{n_j} m(x)x^{-3/2}dx \ll n_j^{1/2+\epsilon},\end{aligned}$$

where $n_j = n_j(T_0)$. Hence,

$$\Sigma_1 \ll V^{-1}(hk)^{-1/2}T^{1/2}Y^{1/2+\epsilon} \ll V^{-1}Y^{1+\epsilon} \ll T^{1+\epsilon}V^{-2}. \tag{5.23}$$

For estimating Σ_2, we use Lemma 3 with $G(x) = ((1 + C_2m/x)(1 + C_2n/x))^{-1/4}$ and $F(x) = 2x(\phi(C_2m/x) - \phi(C_2n/x))$. From the definition (4.6) of ϕ we get

$$\begin{aligned}F'(x) &= 2\,\mathrm{arcsinh}((C_2m/x)^{1/2}) - 2\,\mathrm{arcsinh}((C_2n/x)^{1/2})\\ &\quad + 4((C_2m/x)(1 + C_2m/x))^{1/2} - 4((C_2n/x)(1 + C_2n/x))^{1/2}.\end{aligned}$$

Therefore $G(x)/F'(x)$ is monotonic in the interval $[T/2, T]$, after replacing $F(x)$ by $-F(x)$ if necessary. We also have $|F'(x)/G(x)| \gg (C_2/T)^{1/2}|\sqrt{m} - \sqrt{n}|$. Hence, by Lemma 3 we get that the integral in Σ_2 is $\ll (T/C_2)^{1/2}|\sqrt{m} - \sqrt{n}|^{-1} \ll (hkT)^{1/2}|\sqrt{m} - \sqrt{n}|^{-1}$. Thus we estimate Σ_2 as follows.

$$\begin{aligned}\Sigma_2 &\ll V^{-1}\sum_j \sum_{\substack{m,n<n_j\\ m\ne n}} |w_j(m)\rho(m)w_j(n)\rho(n)|(mn)^{-1/4}|\sqrt{m} - \sqrt{n}|^{-1}\\ &\ll V^{-1}\sum_j \sum_{n<m<n_j} |w_j(n)\rho(n)w_j(m)\rho(m)|n^{-1/4}m^{1/4}(m-n)^{-1}\\ &\ll V^{-1}\sum_j \sum_{h<n_j} h^{-1} \sum_{n<n_j-h} |w_j(n)\rho(n)w_j(n+h)\rho(n+h)|n^{-1/4}(n+h)^{1/4},\end{aligned}$$

where $n_j = n_j(T_0)$. Using the Cauchy-Schwarz inequality, the inner sum over n is bounded by

$$\ll \left(\sum_{n<n_j} |w_j(n)\rho(n)|^2 n^{-1/2}\right)^{1/2} \left(\sum_{n+h<n_j} |w_j(n+h)\rho(n+h)|^2(n+h)^{1/2}\right)^{1/2}.$$

By using (5.22) and partial summation as was done in estimating Σ_1, we conclude that the above is

$$\ll (n_j^{1/2+\epsilon} n_j^{3/2+\epsilon})^{\frac{1}{2}} = n_j^{1+\epsilon}.$$

Hence, we combine this with $\sum_{h<x} h^{-1} \ll \log x$ to finally get the following estimate

$$\Sigma_2 \ll V^{-1} \sum_j n_j^{1+\epsilon} \log n_j \ll V^{-1} Y^{1+\epsilon} \ll T^{1+\epsilon} V^{-2}. \tag{5.24}$$

Thus from (5.23), (5.24) and (5.20), we get

$$R \ll T^{1+\epsilon} V^{-2}.$$

This completes the proof of the theorem.

Acknowledgements

The authors wish to thank the referee for his/her suggestions in an earlier version of this paper.

References

[1] R. Balasubramanian, An improvement of a theorem of Titchmarsh on the mean square of $|\zeta(1/2+it)|$, *Proc. London Math. Soc.*, **36** (1978) 540–576.

[2] P. T. Bateman and E. Grosswald, On Epstein's zeta function, *Acta Arith.*, **9** (1964) 365–373.

[3] Bruce Berndt, The number of zeros of the Dedekind zeta-function on the critical line, *J. Number Theory*, **3** (1971) 1–6.

[4] K. Chandrasekharan and Raghavan Narasimhan, Zeta-functions of ideal classes in quadratic fields and their zeros on the critical line, *Comment. Math. Helv.*, **43** (1968) 18–30.

[5] S. Chowla and A. Selberg, On Epstein's zeta function, *I. Proc. Nat. Acad. Sci. U. S. A.*, **35** (1949) 371–374.

[6] H. Davenport and H. Heilbronn, *J. London Math. Soc.*, **11** (1936) 181–185; ibid., **11** (1936) 307–312.

[7] G. H. Hardy and J. E. Littlewood, Contributions to the theory of the Riemann zeta-function and the distribution of primes, *Acta Math.*, **41** (1918) 119–196.

[8] D. R. Heath-Brown, Simple zeros of the Riemann zeta-function on the critical line, *Bull. London Math. Soc.*, **11** (1979) 17–18.

[9] A. Ivić, The Riemann zeta-function, The theory of the Riemann zeta-function with applications, A Wiley-Interscience Publication. John Wiley & Sons, Inc., New York (1985) xvi+517.

[10] A. Ivić and M. Jutila, Gaps between consecutive zeros of the Riemann zeta-function on the critical line, *Monatsh. Math.*, **105** (1988) no. 1, 59–73.

[11] M. Jutila, Transformation Formulae for Dirichlet Polynomials, *J. Number Theory*, **18** (1984) no. 2, 135–156.

[12] M. Jutila, Lectures on a Method in the Theory of Exponential sums, *Tata institute of Fundamental Research, Lectures on Mathematics and Physics*, **80**; published for the Tata institute of Fundamental Research by Springer-Verlag, Berlin-Heidelberg-New York-Tokyo (1987).

[13] M. Jutila, Exponential Sums Connected with Quadratic Forms, *Number Theory* (ed. Richard A. Mollin), 271–286, Walter de Gruyter & Co, Berlin-New York (1990).

[14] M. Jutila and K. Srinivas, Gaps between the zeros of Epstein's zeta-functions on the critical line, *Bull. London Math. Soc.*, **37** (2005) no. 1, 45–53.

[15] A. A. Karatsuba, On the distance between consecutive zeros of the Riemann zeta-function on the critical line (Russian), *Trudy Mat. Inst. Steklova AN SSSR*, **157** (1981) 49–63.

[16] A. A. Karatsuba, The distribution of zeros of $\zeta\left(\frac{1}{2}+it\right)$ (Russian), *Izv. Akad. Nauk SSSR Ser. Mat.*, **48** (1984) 1214–1224.

[17] J. Moser, On a theorem of Hardy-Littlewood in the theory of the Riemann zeta-function (Russian), *Acta Arith.*, **31** (1976b) 45–51 and **35** (1979) 403–404.

[18] M. Ram Murty and R. Osburn, Representations of integers by certain positive definite binary quadratic forms, *Ramanujan J.*, **14** (2007) no. 3, 351–359.

[19] R. Osburn, A remark on a conjecture of Borwein and Choi, *Proc. Amer. Math. Soc.*, **133** (2005) no. 10, 2903–2909.

[20] H. S. A. Potter and E. C. Titchmarsh, The zeros of Epstein's zeta-functions, *Proc. London Math. Soc.*, **39** (1935) no. 2, 372–384.

[21] K. Ramachandra, On the Mean-Value and Omega-Theorems for the Riemann Zeta-Function, *Tata Inst. Fund. Res. Lect. Math.*, **85** (Springer, Berlin/Heidelberg/New York/Tokyo, 1995).

[22] A. Sankaranarayanan, Zeros of quadratic zeta-functions on the critical line, *Acta Arith.*, **69** (1995) 21–37.

[23] Atle Selberg, On the zeros of Riemann's zeta-function, *Skr. Norske Vid. Akad. Oslo I.*, **10** (1942) 1–59.

[24] R. A. Smith, The average order of a class of arithmetic functions over arithmetic progressions with applications to quadratic forms, *J. Reine Angew. Math.*, **317** (1980) 74–87.

[25] H. M. Stark, On the zeros of Epstein's zeta function, *Mathematika*, **14** (1967) 47–55.

[26] E. C. Titchmarsh, The theory of the Riemann zeta-function, Second edition. Edited and with a preface by D. R. Heath-Brown, The Clarendon Press, Oxford University Press, New York (1986).

[27] S. M. Voronin, On the zeros of zeta-functions of quadratic forms (in Russian), *Trudy Mat. Inst. Steklov*, **142** (1976) 135–147.

TCV Proceedings No. 15, 2011, pp. 89–101.

L-Series and Transcendental Numbers

M. Ram Murty

Department of Mathematics, Queen's University, Kingston, Ontario, K7L 3N6, Canada
e-mail: murty@mast.queensu.ca

Abstract. We give a survey of some recent work on the transcendence of special values of certain abelian and non-abelian L-series.

Keywords. Class group L-functions, Chowla-Milnor conjecture, applications of Baker's theorem, Nesterenko's theorem.

In Honour of Professor Vasudevan's 60th Birthday

1. Introduction

This paper is a survey of some recent work of mine done jointly with V. Kumar Murty, N. Saradha, S. Gun and P. Rath. In all of these works, the motivating question is the following. Given an automorphic representation π, we consider the L-series $L(s, \pi)$ attached to π according to the Langlands formalism. We are interested in the possible transcendence of special values $L(k, \pi)$ when k is a positive integer. Sometimes, we will be interested in algebraic linear combinations of such values. There are many conjectures in the literature regarding the arithmetic significance of $L(k, \pi)$ for certain classes of π, but in this paper, we are concerned with its transcendence.

In these investigations, a central role is played by the celebrated theorem of Alan Baker, proved in 1966. This theorem states the following. If $\alpha_1, \ldots, \alpha_n$ are non-zero algebraic numbers and $\beta_1, \ldots, \beta_n$ are algebraic numbers, then

$$\beta_1 \log \alpha_1 + \cdots + \beta_n \log \alpha_n$$

is either zero or transcendental. The latter case arises if $\log \alpha_1, \ldots, \log \alpha_n$ are linearly independent over $\mathbb{Q}$ and $\beta_1, \ldots, \beta_n$ are not all zero. The latter case also arises if $\beta_1, \ldots, \beta_n$ are linearly independent over $\mathbb{Q}$ (see for example, [20]).

2. Dirichlet's formula

The simplest case of study arises when χ is a non-trivial Dirichlet character mod q and $s = 1$. In this case, one can prove that (see for example, [10])

$$L(1, \chi) = \sum_{a=1}^{q-1} c_a \log(1 - \zeta^a),$$

where the c_a's are algebraic numbers and ζ is a primitive q-th root of unity. Baker's theorem implies that this number is either zero or transcendental. But the well-known theorem of Dirichlet that $L(1,\chi) \neq 0$ now implies that $L(1,\chi)$ is transcendental. This was observed in [1].

3. Class group L-functions

The next "simplest" case to consider is that of Hecke characters of an imaginary quadratic field K. It is convenient to consider a slightly more general context. Let H_K be the ideal class group of K and $f : H_k \to \bar{\mathbb{Q}}$ be an algebraic valued function on H_K not identically zero. As shown in [9], the series

$$\sum_{\mathfrak{a}} \frac{f(\mathfrak{a})}{N(\mathfrak{a})^s}$$

extends to an entire function if and only if

$$\sum_{\mathfrak{a} \in H_K} f(\mathfrak{a}) = 0.$$

In the latter case, we show that $L(1, f)/\pi$ is transcendental. In particular, it is non-zero. This allows us to deduce the following result: the values $L(1,\chi)$ are linearly independent over $\bar{\mathbb{Q}}$ as χ ranges over the characters of H_K. Moreover, all of these numbers are transcendental with at most one exception.

As a consequence of these investigations, V. Kumar Murty and I showed that if χ_D is the quadratic character corresponding to the imaginary quadratic field K, then π and

$$\exp\left(\frac{L'(1,\chi_D)}{L(1,\chi_D)} - \gamma\right),$$

where γ is Euler's constant, are algebraically independent. In particular, the latter exponential is transcendental.

This result raises the interesting question of when we can have $L'(1,\chi) = 0$ for such quadratic characters. For various reasons, to be discussed below, it seems reasonable to conjecture that this never happens for Dirichlet L-series. In this direction, V. Kumar Murty and I proved that for q prime, the number of characters (mod q) for which $L'(1,\chi) = 0$ is $O(q^\epsilon)$ for any $\epsilon > 0$.

The question of non-vanishing of $L'(1,\chi)$ is related to a conjecture due to Lang and Rohrlich. This conjecture predicts that the numbers $\log \Gamma(a/q)$ with $0 < a < q$ and $(a, q) = 1$ are linearly independent over $\bar{\mathbb{Q}}$. Assuming this conjecture, S. Gun, P. Rath and I showed that $L'(1,\chi) \neq 0$ with at most one exceptional character χ mod q.

A weaker conjecture is the assertion that the above numbers are linearly independent over $\mathbb{Q}$. At present, this weaker conjecture is also unknown. However, in the case $q = 3, 4$ and 6, S. Gun, P. Rath and I were able to show that this weaker conjecture is true. This result is an attractive application of a theorem of Chudnovsky (see p. 308 of [6]) and since the proof is short, we give it in the case of $q = 3$ (with the cases $q = 4$ and $q = 6$ being similar).

Suppose that $\log \Gamma(1/3)$ and $\log \Gamma(2/3)$ are linearly dependent over $\mathbb{Q}$. Then,

$$\Gamma(1/3)^a \Gamma(2/3)^b = 1$$

for some integers a, b not both zero. But by the well-known functional equation

$$\Gamma(z)\Gamma(1-z) = \frac{\pi}{\sin \pi z},$$

we have

$$\Gamma(1/3)\Gamma(2/3) = \frac{2\pi}{\sqrt{3}}.$$

Thus,

$$\Gamma(1/3)^{a-b}(2\pi/\sqrt{3})^b = 1.$$

But this contradicts a 1976 result of Chudnovsky (see p. 308 of [6]) that π and $\Gamma(1/3)$ are algebraically independent.

4. The Chowla and Erdös conjectures

In 1965, Erdös (see [8]) conjectured that if f is a function defined on the residue classes mod q with $f(q) = 0$ and $f(a) = \pm 1$ otherwise, then

$$\sum_{n=1}^{\infty} \frac{f(n)}{n} \neq 0.$$

It is not hard to show that the conjecture is true if q is even. In recent work with N. Saradha, I was able to show that the conjecture is true if $q \equiv 3 \pmod 4$.

There is a related conjecture of Chowla [4]. In 1969, Chowla asked at the Stony Brook conference if there is a rational valued function, not identically zero, defined on the residue classes mod p such that

$$\sum_{n=1}^{\infty} \frac{f(n)}{n} = 0.$$

In a beautiful paper by Baker, Birch and Wirsing [2], this question was answered in the negative using Baker's theory of linear forms in logarithms. They studied the case of the general modulus q instead of the prime modulus.

5. Schanuel's conjecture

The celebrated Lindemann-Weierstrass theorem states that if $\alpha_1, \ldots, \alpha_n$ are algebraic numbers which are linearly independent over $\mathbb{Q}$, then $e^{\alpha_1}, \ldots, e^{\alpha_n}$ are algebraically independent. The conjecture of Schanuel asserts that the algebraicity hypothesis can be dropped. Thus, the conjecture says (as cited in Lang's monograph on Transcendental Numbers) that if $x_1, \ldots, x_n$ are linearly independent over $\mathbb{Q}$, then

$$\operatorname{tr\,deg} \mathbb{Q}(x_1, \ldots, x_n, e^{x_1}, \ldots, e^{x_n}) \geq n.$$

This is still a major unsolved problem and many interesting consequences emerge from it. For example, we immediately deduce that e and π are algebraically independent. To see this, consider $x_1 = 1, x_2 = \pi i$. These are linearly independent over $\mathbb{Q}$ and so the transcendence degree of

$$\mathbb{Q}(\pi, e) = \mathbb{Q}(1, \pi i, e, e^{\pi i})$$

is at least 2. Thus, e and π are algebraically independent. Consequently, Schanuel implies that both $e+\pi$ and $e\pi$ are transcendental, for otherwise, we have an algebraic relation between them. In particular, 1 and $\log \pi$ are linearly independent over $\mathbb{Q}$ for if not, we have $a + b \log \pi = 0$, for some integers a, b from which we get $e^a \pi^b = 1$, contradicting the algebraic independence of e and π. To deduce that $\log \pi$ is transcendental, we consider $\pi i, \log \pi$ which are linearly independent over $\mathbb{Q}$ since $\pi \neq \pm 1$. Thus, π and $\log \pi$ are algebraically independent. In particular, $\log \pi$ is transcendental. Many interesting results can be deduced from this conjecture.

One particular result that is relevant to our discussion below is a strengthening of Baker's theorem. Recall that Baker's theorem says that if $\log \alpha_1, \ldots, \log \alpha_n$ are logarithms of algebraic numbers linearly independent over $\mathbb{Q}$, then they are linearly independent over $\bar{\mathbb{Q}}$. Schanuel's conjecture predicts the algebraic independence of these numbers. We refer to this as the weak Schanuel conjecture since it plays a major role in our discussion below.

6. Artin L-series

Our motivating question in the context of Artin L-series can be formulated precisely in the following way. If χ is an Artin character, which does not contain the trvial character, then is it true that $L(1, \chi)$ is transcendental? It seems reasonable to conjecture that the answer is yes, thus giving a non-abelian generalization of the result for Dirichlet characters that was discussed at the outset.

In this context, Stark [19] has made specific conjectures about the special value $L(1, \chi)$ as the determinant of a regulator-like matrix involving units of number fields. If χ is a rational character, Stark was able to prove his conjecture.

In joint work with S. Gun and P. Rath, I [13] was able to show that if we admit Stark's conjecture, along with the weak Schanuel conjecture, then one can deduce that $L(1, \chi)$ is transcendental. If χ is a rational character, we can drop the assumption of Stark's conjecture since that is known. These are first steps towards resolving the general conjecture that $L(1, \chi)$ is transcendental for any Artin character χ.

7. Catalan's constant

The study of $L(s, \chi)$ at $s = 1$ suggests an investigation of other values like $L(k, \chi)$ where k is a positive integer greater than 1. Even in the case of classical Dirichlet characters, many things are still shrouded in mystery and inevitably lead to questions as difficult as the transcendence of $\zeta(3)$.

If χ is a Dirichlet character (mod q), so that χ and k have the same parity, then one can show, following Euler, that $L(k, \chi)$ is a non-zero algebraic multiple of π^k. Consequently, $L(k, \chi)$ is transcendental in these cases. If k and χ have opposite parity, then the arithmetic study of $L(k, \chi)$ seems impenetrable at the moment.

The "simplest" case arises if $q = 4$ and χ is the (odd) non-trivial character mod 4. In this case, the number $L(2, \chi)$ is called Catalan's constant and it is unknown at present if it is irrational. In 2003, Rivoal and Zudilin [17] showed that at least one of the seven numbers, $L(2k, \chi)$, with $k = 1, 2, \ldots, 7$, is irrational.

Part of the difficulty is exposed when one attempts at an explicit evaluation of $L(2, \chi)$. This involves Barnes double gamma function so that the transcendence of Catalan's constant is intimately tied up with the study of special values of the classical gamma function and the Barnes double gamma function.

In 1900, Barnes [3] introduced the double gamma function $G(z)$. It has a Weierstrass factorization:

$$G(z+1) = (2\pi)^z e^{-z(z+1)/2+\gamma z^2/2} \prod_{n=1}^{\infty} \left(1 + \frac{z}{n}\right)^n e^{-z+z^2/(2n)}.$$

One can show that

$$2L(2, \chi)/\pi = 3 \log \Gamma(1/4) - \log \Gamma(3/4) 4 \log G(1/4) - 4 \log G(3/4).$$

S. Gun, P. Rath and myself [14] were able to show that at least one of the numbers

$$L(2, \chi)/\pi^2, \quad G(1/4)/G(3/4),$$

is transcendental. Presumably, both numbers are transcendental.

8. Multiple gamma functions and $\zeta(3)$

A similar result can be proved for $\zeta(3)/\pi^3$ by relating it to multiple gamma functions. These functions were also introduced by Barnes and are defined inductively as follows. Let $\Gamma_1(z)$ be the classical gamma function. The multiple gamma functions Γ_n satisfy the functional equation

$$\Gamma_{n+1}(z+1) = \frac{\Gamma_{n+1}(z)}{\Gamma_n(z)},$$

with the initial condition $\Gamma_n(1) = 1$. $\Gamma_2(z)$ is Barnes double gamma function.

With S. Gun and P. Rath, I [14] proved that at least one of

$$\zeta(3)/\pi^3, \quad \Gamma_2(1/2)\Gamma_3(1/2),$$

is transcendental.

9. Chowla's problem revisited

As mentioned earlier, in 1969, Chowla asked if there is a non-zero rational valued function f with prime period such that

$$\sum_{n=1}^{\infty} \frac{f(n)}{n} = 0.$$

In 1973, Baker, Birch and Wirsing [2] answered Chowla's question by proving the following. Let f be an algebraic valued function, not identically zero, defined on the integers with period q (not necessarily prime). Suppose that $f(r) = 0$ for $1 < (r, q) < q$ and the q-th cyclotomic polynomial is irreducible over the field $\mathbb{Q}(f(n) : n \in \mathbb{Z})$. Then,

$$\sum_{n=1}^{\infty} \frac{f(n)}{n} \neq 0.$$

The proof of this theorem uses Baker's theory of linear forms in logarithms.

In the setting of Chowla's problem, the series

$$\sum_{n=1}^{\infty} \frac{f(n)}{n}$$

can be written as a linear form in logarithms. Consequently, for any f satisfying the conditions of the theorem of Baker, Birch and Wirsing, we deduce that the series is transcendental.

In the special case f is a Dirichlet character χ, we combine the classical theorem of Dirichlet that $L(1, \chi)$ in not zero along with Baker's theorem to deduce that it is transcendental. This was observed in [1]. If $(q, \varphi(q)) = 1$, then the Baker–Birch–Wirsing theorem allows us to deduce that the $L(1, \chi)$ as χ ranges over the non-trivial characters (mod q) are linearly independent over the rationals.

As noted in [10], these results allow us to deduce something new about the digamma function. Recall that the digamma function, $\psi(x)$ is the logarithmic derivative of the gamma function. Its connection to Chowla's problem is revealed via Hurwitz's zeta function:

$$\zeta(s, x) = \sum_{n=0}^{\infty} \frac{1}{(n+x)^s}.$$

This series admits an analytic continuation to the entire complex plane apart from a simple pole at $s = 1$ with residue 1. In fact, one can show that

$$\zeta(s, x) = \frac{1}{s-1} - \psi(x) + O(s-1),$$

as $s \to 1$.

The analytic continuation of $L(s, f)$ can be determined from the analytic continuation of the Hurwitz zeta function. More precisely,

$$L(s, f) = q^{-s} \sum_{a=1}^{q} f(a)\zeta(s, a/q).$$

Using this, one can show that

$$\sum_{n=1}^{\infty} \frac{f(n)}{n}$$

converges if and only if

$$\sum_{a=1}^{q} f(a) = 0.$$

In the case of convergence, we have

$$L(1, f) = -\frac{1}{q} \sum_{a=1}^{q} f(a)\psi(a/q). \tag{1}$$

If ζ denotes a primitive q-th root of unity, we can write

$$f(n) = \sum_{a=1}^{q} \hat{f}(a)\zeta^{an}.$$

Here

$$\hat{f}(a) = \frac{1}{q} \sum_{b=1}^{q} f(b)\zeta^{-ab}.$$

Using this, it is not hard to see that

$$L(1, f) = -\sum_{a=1}^{q-1} \hat{f}(a) \log(1 - \zeta^{a}).$$

Clearly, if f is algebraic valued, so is $\hat{f}$. Consequently, the right hand side is by the theorem of Baker, zero or transcendental. Thus, if $L(1, f) \neq 0$, then it is transcendental.

The above connection with the Hurwitz zeta function allows us to deduce some new results about the digamma function. For instance, we can show that there is at most one algebraic number in the list $\psi(a/q)$ with $0 < a < q$ and $(a, q) = 1$. Indeed, if both $\psi(a_1/q)$ and $\psi(a_2/q)$ are algebraic, then choosing $f(a_1) = 1$, $f(a_2) = -1$ and $f(a) = 0$ for all the other classes, we see that the conditions of the Baker, Birch and Wirsing theorem are satisfied. So $L(1, f) \neq 0$ and transcendental. On the other hand, from formula (1), we see that $L(1, f)$ is algebraic. This is a contradiction.

10. Euler constants

In 1975, Lehmer [7] defined generalized Euler constants as follows:

$$\gamma(a, q) = \lim_{x \to \infty} \left(\sum_{n \leq x, n \equiv a (\mathrm{mod}\ q)} \frac{1}{n} - \frac{\log x}{q} \right).$$

In the case $q = 1$, this is Euler's constant γ. One expects that all of these generalized Euler constants are transcendental. This is unknown at the moment. N. Saradha and I [11] showed that in the infinite list

$$\gamma, \quad \gamma(a, q), \quad 1 \leq a < q, \quad q \geq 2,$$

there is at most one algebraic number.

11. The Chowla–Milnor conjecture

So far, we have discussed certain L-series at $s = 1$. Now we shall discuss L-series at other positive integer arguments. In 1982, P. and S. Chowla [5] made a conjecture about $L(s, f)$ when f is a rational-valued function defined on the residue classes mod q. When q is prime, they predicted that $L(2, f) \neq 0$ unless $f(1) = f(2) = \cdots = f(p-1) = f(p)/(1-p^2)$. This conjecture is equivalent to the following: the numbers

$$\zeta(2, a/p), \quad 1 \leq a \leq p-1$$

are linearly independent over $\mathbb{Q}$. Inspired by this conjecture, Milnor [15] suggested the following generalization; the numbers

$$\zeta(k, a/q), \quad 1 \leq a \leq q, \quad (a, q) = 1,$$

are linearly independent over $\mathbb{Q}$. We refer to this assertion as the Chowla–Milnor conjecture.

This innocuous looking conjecture has serious implications. In joint work with S. Gun and P. Rath, I [14] showed the following. The Chowla–Milnor conjecture for $q = 4$ is equivalent to the irrationality of

$$\zeta(2k+1)/\pi^{2k+1}$$

for all $k \geq 1$. Unconditionally, we can show that the Chowla–Milnor conjecture is true for k odd and $q = 3$ or $q = 4$.

Our investigations led us to formulate the strong Chowla–Milnor conjecture. This is the following assertion. For any k and $q > 1$, the numbers 1 and $\zeta(k, a/q)$, $1 \leq a < q$ with $(a, q) = 1$ are linearly independent over $\mathbb{Q}$. In joint work with S. Gun and P. Rath, I [14] showed that for k odd, the strong Chowla conjecture is true for either $q = 3$ or for $q = 4$ if and only if $\zeta(k)$ is irrational.

12. The polylog conjecture

The polylogarithm function $Li_k(z)$ is defined for $|z| < 1$ by the series

$$Li_k(z) = \sum_{n=1}^{\infty} \frac{z^n}{n^k}.$$

For $k > 1$, we have convergence for $|z| \leq 1$. Let us note that $Li_1(z) = -\log(1-z)$. It seems reasonable to conjecture the following. Suppose that $\alpha_1, \ldots, \alpha_n$ are algebraic numbers such that $Li_k(\alpha_1), \ldots, Li_k(\alpha_n)$ are linearly independent over $\mathbb{Q}$, then they are linearly independent over $\bar{\mathbb{Q}}$. In the case $k = 1$, this is a consequence of the celebrated theorem of Alan Baker.

With S. Gun and P. Rath, I [14] proved that the polylog conjecture implies the Chowla–Milnor conjecture for all k and all q.

13. The Chowla–Milnor space

Let $V_k(q)$ be the $\mathbb{Q}$-vector space spanned by the numbers $\zeta(k, a/q)$ with $1 \leq a < q$, and $(a, q) = 1$. The Chowla–Milnor conjecture is equivalent to the assertion that $\dim V_k(q) = \varphi(q)$.

S. Gun, P. Rath and I [14] were able to show the following. Let $k > 1$ and $q > 1$. Then,

$$\dim_{\mathbb{Q}} V_k(q) \geq \varphi(q)/2.$$

If k is odd and r is coprime to q, then either

$$\dim_{\mathbb{Q}} V_k(q) \geq \frac{\varphi(q)}{2} + 1,$$

or

$$\dim_{\mathbb{Q}} V_k(r) \geq \frac{\varphi(r)}{2} + 1.$$

In particular, there exists a q_0 such that

$$\dim_{\mathbb{Q}} V_k(q) \geq \frac{\varphi(q)}{2} + 1,$$

for any q coprime to q_0.

14. Multiple zeta values

There is a surprising connection between the theory of multiple zeta values and the Chowla–Milnor conjecture that emerged in our researches. To explain this, we begin with some definitions.

The multiple zeta function is defined by

$$\zeta(s_1, \ldots, s_k) = \sum_{n_1 > n_2 > \cdots > n_k \geq 1} \frac{1}{n_1^{s_1} n_2^{s_2} \cdots n_k^{s_k}}.$$

We usually consider this function with the s_i's positive integers and with $s_1 > 1$ for convergence reasons. Let V_k be the $\mathbb{Q}$-vector space spanned by the values $\zeta(s_1, \ldots, s_t)$ with $t \geq 1$, $s_1 > 1$ and $s_1 + \cdots + s_t = k$. Let d_k be the dimension of V_k.

Zagier [21] has conjectured that d_k satisfies the Fibonacci-like recurrence

$$d_k = d_{k-1} + d_{k-3}.$$

The recurrence is easily analysed with initial conditions, $d_0 = 1$, $d_1 = 0$ and $d_2 = 1$. It is apparent that d_k has exponential growth.

If we define $\delta_0 = 1, \delta_1 = 0, \delta_2 = 1$ and

$$\delta_k = \delta_{k-2} + \delta_{k-3},$$

then Deligne, Goncharov and Terasoma (independently) showed that $d_k \leq \delta_k$ for every value of k. However, to date, no example of any value of k is known for which $d_k \geq 2$.

It is not hard to see that $d_2 = d_3 = d_4 = 1$. It is also not difficult to show that $d_5 = 2$ if and only if $\zeta(2)\zeta(3)/\zeta(5)$ is irrational. Similarly, one can show that $d_6 = 2$ if and only if $\zeta^2(3)/\pi^6$ is irrational.

In joint work with S. Gun and P. Rath, I [14] was able to show that if the Chowla–Milnor conjecture is true for all values of q and k, then

$$\left(\frac{\zeta(2k+1)}{\pi^{2k+1}}\right)^2$$

is irrational and $d_{4k+2} \geq 2$ for all $k \geq 1$.

In 1948, Petersson [16] proved that $7\zeta(3)/\pi^3 = (f, g)_\Gamma$ where Γ is the theta subgroup of $SL_2(\mathbb{Z})$. This raises the interesting question of transcendental values of the Petersson inner product.

In joint work with S. Gun and P. Rath, I [13] was able to show the following. Let f be a non-zero normalized Hecke eigenform of weight 1 and level N. Suppose that f has rational Fourier coeffficients. Then, assuming the weak Schanuel conjecture, the inner product (f, f) is transcendental.

15. Kronecker's limit formula

Let K be an imaginary quadratic field and H_K its ideal class group. For each ideal class C of H_K, we have the ideal class zeta function, $\zeta(s, C)$. Hecke showed how to analytically continue this zeta function to the entire complex plane, apart from a simple pole at $s = 1$. Kronecker's celebrated limit formula (see for example, [18]) states that

$$\lim_{s\to 1} \zeta(s, C) = \frac{2\pi}{w\sqrt{|d_K|}}\left(\frac{1}{s-1} + 2\gamma - \log|d_K| - \frac{1}{6}\log g(C^{-1})\right) + O(s-1),$$

where d_K is the discriminant of K and w is the number of roots of unity in K. Here g is defined as follows. If $\mathfrak{a}$ is an ideal with basis u, v where $\Im(u/v) > 0$, we define

$$g(\mathfrak{a}) = (2\pi)^{-12} N(\mathfrak{a})^6 |\Delta(u/v)|,$$

where $\Delta(z)$ is Ramanujan's cusp form.

One can formulate a K-analogue of Chowla's problem. Let $f : H_K \to \bar{\mathbb{Q}}$ be not identically zero. Set

$$L(s, f) = \sum_{\mathfrak{a}} \frac{f(\mathfrak{a})}{N(\mathfrak{a})^s}.$$

Using the theory of the ideal class zeta function, Kumar Murty and I [9] showed that $L(s, f)$ converges at $s = 1$ if and only if

$$\sum_{C \in H_K} f(C) = 0.$$

In the latter case, one can prove that $L(1, f) \neq 0$ and

$$-\frac{L(1, f)}{\pi} = \frac{1}{3w\sqrt{|d_K|}} \sum_{C \in H_K} f(C) \log g(C^{-1}).$$

This formula is not in a suitable form to apply Baker's theorem. To put it in an amenable form, we apply the theory of complex multiplication. This theory says that if C_1 and C_2 are two ideal classes, then $g(C_1)/g(C_2)$ is an algebraic number lying in the Hilbert class field of K. Since

$$\sum_{C \in H_K} f(C) = 0,$$

we can re-write our expression as

$$-\frac{L(1, f)}{\pi} = \frac{1}{3w\sqrt{|d_K|}} \sum_{C \in H_K} f(C) \log g(C^{-1})/g(C_0).$$

The right hand side is now a $\bar{\mathbb{Q}}$-linear combination of logarithms of algebraic numbers. Since $L(1, f) \neq 0$, we deduce from Baker's theorem that $L(1, f)/\pi$ is transcendental.

16. The Chowla–Selberg formula

Applying the previous discussion to the case of an ideal class character χ, we deduce a formula for $L(1, \chi)$ and infer that $L(1, \chi)/\pi$ is transcendental.

In the particular case that χ is a genus character, we can deduce more. Indeed, if χ_D is the character corresponding to the imaginary quadratic field K, we deduce

$$\frac{L'(1, \chi_D)}{L(1, \chi_D)} = \gamma - \log |d_K| - \frac{1}{6h} \sum_{C \in H_K} \log g(C),$$

where h denotes the order of H_K.

This allows us to give a new proof of the Chowla–Selberg formula. Indeed, by the functional equation, we can relate

$$\frac{L'(1, \chi_D)}{L(1, \chi_D)} \quad \text{with} \quad \frac{L'(0, \chi_D)}{L(0, \chi_D)}.$$

On the other hand,

$$L(s, \chi_D) = D^{-s} \sum_{a=1}^{D} \chi_D(a) \zeta(s, a/D).$$

Applying the formula of Lerch (see for example, [15])

$$\zeta'(0, x) = \log(\Gamma(x)/\sqrt{2\pi}),$$

as well as the class number formula for $L(0, \chi_D)$, we deduce the celebrated Chowla–Selberg formula.

Putting everything together, we get

$$\prod_{C \in H_K} g(C)^6 = \left(\frac{1}{4\pi |d_K|}\right)^h \prod_{a=1}^{D} \Gamma(a/D)^{w\chi_D(a)/2}.$$

By the work of Deligne-Gross, the left hand side is (up to an algebraic factor) the period of an elliptic curve with CM by the ring of integers of K. In other words,

$$\prod_{a=1}^{D} \Gamma(a/D)^{\chi_D(a)h},$$

is a product of a power of π and a power of a period of a CM elliptic curve (up to an algebraic factor).

17. Transcendental values of the Γ-function

The preceding analysis can be used to obtain new transcendental values of the Γ-function. If q is a divisor of 24, it is not hard to see that all the characters (mod q) are real-valued. Thus, if $q|24$, then $\Gamma(a/q)$ can be expressed as a product of periods of non-isogenous CM elliptic curves and π. Following Grothendieck, it seems reasonable to conjecture that the fundamental periods of non-isogenous CM elliptic curves and π are algebraically independent. If we admit this conjecture, then we can deduce that $\Gamma(a/q)$ with $q|24$ is transcendental. In particular, $\Gamma(1/8)$ will be transcendental.

It is possible to deduce something unconditionally. We can show that at most one of $\Gamma(1/8), \Gamma(3/8), \Gamma(5/8), \Gamma(7/8)$ is algebraic. Here is a sketch of the proof. Suppose that at least two of these numbers are algebraic. By the Chowla–Selberg formula, we get an algebraic dependency relation between π and a period of a CM elliptic curve. This contradicts a theorem of Chudnovsky that states that these numbers are algebraically independent.

Acknowledgements

I am grateful to Drs. Sanoli Gun and Purusottam Rath for their corrections of an earlier version of this paper. I also thank the referee for helpful remarks.

References

[1] S. Adhikari, N. Saradha, T. N. Shorey and R. Tijdeman, Transcendental infinite sums, *Indag. Math. (N.S.)*, **12** (2001) no. 1, 1–14.

[2] A. Baker, B. Birch and E. A. Wirsing, On a problem of Chowla, *Journal of Number Theory*, **5** (1973) 224–236.

[3] E. W. Barnes, The theory of the G-function, *Quarterly Journal of Pure and Applied Math.*, **31** (1900) 264–314.

[4] S. Chowla, The non-existence of non-trivial relations between roots of a certain irreducible equation, *Journal of Number Theory*, **2** (1970) 120–123.

[5] P. Chowla and S. Chowla, On irrational numbers, *K. Norske Vidensk. Selsk. Skr.*, **3** (1982) 1–5. (See also: The Collected Papers of Sarvadaman Chowla, Volume 3, 1962–1986, edited by J. G. Huard and K. S. Williams, Centre de Recherches Mathématiques, 1999, pp. 1383–1387.)

[6] G. V. Chudnovsky, Contributions to the theory of transcendental numbers, Mathematical Surveys and Monographs, Number 19, Amer. Math. Society, Providence, Rhode Island (1984).

[7] D. H. Lehmer, Euler constants for arithmetical progressions, *Acta Arithmetica*, **27** (1975) 125–142.

[8] A. E. Livingston, The series $\sum_1^\infty f(n)/n$ for periodic f, *Canadian Math. Bulletin*, **8** (1965) 413–432.

[9] M. Ram Murty and V. Kumar Murty, Transcendental values of class group L-functions, *Mathematische Annalen*, published online December 12 (2010).

[10] M. Ram Murty and N. Saradha, Transcendental values of the digamma function, *Journal of Number Theory*, **125** (2007) 298–318.

[11] M. Ram Murty and N. Saradha, Euler-Lehmer constants and a conjecture of Erdös, *Journal of Number Theory*, **130** (2010) no. 12, 2671–2682.

[12] M. Ram Murty and N. Saradha, Special values of the polygamma functions, *International Journal of Number Theory*, **5** (2009) no. 2, 257–270.

[13] S. Gun, M. Ram Murty and P. Rath, Transcendental values of special values of L-series, *Canadian Journal of Mathematics*, published online November 6 (2010).

[14] S. Gun, M. Ram Murty and P. Rath, On a conjecture of Chowla and Milnor, to appear in *Canadian Journal of Mathematics*.

[15] J. Milnor, On polylogarithms, Hurwitz zeta functions and the Kubert identities, *Enseign. Math.*, **29(2)** (1983) no. 3–4, 281–322.

[16] H. Petersson, Über die Berechnung der Skalarprodukte ganzer Modulformen, *Comm. Math. Helvetici*, **22** (1983) no. 1, 168–199.

[17] T. Rivoal and W. Zudilin, Diophantine properties of numbers related to Catalan's constant, *Math. Annalen*, **326** (2003) 705–721.

[18] C. L. Siegel, Advanced analytic number theory, TIFR Publications, Tata Press (1980).

[19] H. Stark, L-functions at $s = 1$, II, Artin L-functions with rational characters, *Advances in Math.*, **17** (1975) no. 1, 60–92.

[20] R. Tijdeman, Hilbert's seventh problem: on the Gel'fond-Baker method and its applications, in Mathematical developments arising from Hilbert problems, *Proc. Symp. Pure Math.* (1974) 241–268, Amer. Math. Society, Providence, Rhode Island (1976).

[21] D. Zagier, Values of zeta functions and their applications, First European Congress of Mathematics, Vol. 2 (Paris, 1992), Progress in Mathematics, **120**, Birkhäuser (1994).

TCV Proceedings No. 15, 2011, pp. 103–111.

The Field of Fourier Coefficients of a Modular Form

V. Kumar Murty

Department of Mathematics, University of Toronto, 40 St. George Street, Toronto, Canada, M5S 2E4
e-mail: murty@math.toronto.edu

Abstract. Given a cusp form f of squarefree level and trivial character, which is a normalized eigenform for the Hecke operators, we determine effective bounds $B = B(f)$ so that for some prime $p < B$, the pth Fourier coefficient generates the field E_f generated by the Fourier coefficients of f. We also show that most primes have this property, in the sense that for a set of primes p of density 1, the pth Fourier coefficient generates E_f. We give effective estimates for the number of primes $\leq x$ that fail to satisfy this property. In the case that the form f has weight 2, we interpret these results in terms of the Abelian variety associated to f.

Keywords. Modular form, Fourier coefficients, Hecke operators.

To Professor T. C. Vasudevan with respect and friendship

1. Introduction

Let N be a positive integer and set as usual

$$\Gamma_1(N) = \left\{ g \in \mathrm{SL}_2(\mathbb{Z}) : g \equiv \begin{pmatrix} 1 & * \\ 0 & 1 \end{pmatrix} \pmod{\mathrm{N}} \right\}.$$

Let f be a holomorphic cusp form for $\Gamma_1(N)$ of weight k and character ϵ and suppose that it is a normalized eigenform for the Hecke operators. Let us write

$$f(z) = \sum_{n=1}^{\infty} a_f(n) e^{2\pi i n z}$$

for its Fourier expansion at the cusp $i\infty$ with $a_f(1) = 1$. In this case, it is known that the Fourier coefficients $a_f(n)$ are algebraic and that they generate a number field, E_f (say).

We are interested in the question of how many Fourier coefficients are actually needed to generate E_f. Suppose that the level N is squarefree and that $\epsilon = 1$. Under these hypotheses, as we shall see, the answer is that a single Fourier coefficient is sufficient to generate this field. We give effective and explicit bounds on the size of such a prime. If we do not make any assumption about N and ϵ, we can still show that

Research partially supported by a Discovery Grant from NSERC.

there is a distinguished subfield F_f of E_f (defined in terms of "inner twists" of f) that is generated by a single Fourier coefficient.

After this paper was written, Vijay Patankar pointed out [1] (which at the time was in preprint form). In that paper, the authors show that under the same conditions as given here, a single Fourier coefficient generates the field of Fourier coefficients. However, their results are ineffective.

2. The field of Fourier coefficients

It is well-known that there is a basis for the space $S_k(N)$ of cusp forms of weight k, trivial character and level N having rational Fourier coefficients. Moreover, it is also known that if the first $L_{k,N}$ Fourier coefficients of such a form vanish, then the form is identically zero, where $L_{k,N} = \frac{k}{12} \cdot [SL_2(\mathbb{Z}) : \Gamma_0(N)]$. From these two facts, it follows that the first $L_{k,N}$ Fourier coefficients determine the form and, in particular, the field generated by the Fourier coefficients.

The statement that a form whose first $L_{k,N}$ coefficients vanish is identically zero can be restated as follows. If two forms $f, g \in S_k(N)$ have the property that $a_f(n) = a_g(n)$ for all $n \leq L_{k,N}$, then $f = g$. To prove our result, we need the following. We would like to find an n for which the collection $\{a_f^\sigma(n)\}$ are all distinct. Here, σ ranges over the various embeddings Σ of E_f into $\mathbb{C}$. This requires the use of the ℓ-adic representations associated to f and the effective Chebotarev density theorem.

We remark that given an elliptic curve $\mathcal{E}$ defined over $\mathbb{Q}$, it is now known that there is a modular form f of weight 2 with rational integer Fourier coefficients for which

$$|\mathcal{E}(\mathbb{F}_p)| = p + 1 - a_f(p)$$

for all but finitely many primes p. Since there are infinitely many elliptic curves, it follows that there are infinitely many f of weight 2 with $E_f = \mathbb{Q}$. As far as I am aware, there is no proof of this without resorting to the deep theorem stated above about elliptic curves.

Before closing this section, we remark that the questions considered here are also of interest in the case of modular forms of half-integral weight. In particular, the problem of finding an effective bound for the least n for which the nth Fourier coefficent of a non-zero modular form is not zero was considered by Manickam, Ramakrishnan and the author in [2]. We make further remarks on this in the final section.

3. Statement of results

We shall work with non-CM forms of squarefree level and trivial character. A form f is of CM type if there is a Dirichlet character χ_f with the property that for all but finitely many primes p, we have

$$a_f(p) = a_f(p)\chi_f(p).$$

In other words, we require that $a_f(p) = 0$ whenever $\chi_f(p) \neq 1$. We say that f is a non-CM form if it is not of CM type. Our first result shows that for such a form, a single Fourier coefficient suffices to determine the field E_f. More precisely, we prove the following.

Theorem 3.1. *Suppose that* N *is squarefree,* $\epsilon = 1$ *and that* f *is a normalized eigenform of the Hecke operators that is not of CM-type. Assume the Generalized Riemann Hypothesis (GRH). Then, for a set of primes* p *of density* 1, *we have*

$$E_f = \mathbb{Q}(a_f(p)). \tag{1}$$

In other words, this single Fourier coefficient generates the entire field of Fourier coefficients. The number of primes $p \leq x$ *for which* (1) *fails to hold is* $\mathbf{O}(x^{13/14})$.

We remark that it is possible to prove an unconditional version of the above result. In this case, we find that (1) holds apart from

$$\ll \frac{x(\log\log x)^2}{(\log x)^{22/21}} \tag{2}$$

of the primes $p \leq x$.

Next, we consider the question of finding a bound for the least prime p for which $E_f = \mathbb{Q}(a_f(p))$. To state these results, we need some discussion of the Galois representations associated to f.

4. Galois representations

We have a family of Galois representations associated to f,

$$\rho_{f,\ell} : \mathrm{Gal}(\bar{\mathbb{Q}}/\mathbb{Q}) \longrightarrow \mathrm{GL}_2(\mathcal{O}_{E_f} \otimes \mathbb{Z}_\ell).$$

Each $\rho_{f,\ell}$ is a semisimple representation, unramified outside primes dividing ℓN and at an unramified prime p, the characteristic polynomial of Frobenius is given by

$$T^2 - a_f(p)T + \epsilon(p)p^{k-1}.$$

The image of $\rho_{f,\ell}$ has been extensively studied. We recall some results of Momose and Ribet that will be used in the sequel. We consider embeddings $\sigma : E_f \hookrightarrow \mathbb{C}$ and the corresponding conjugate form f^σ with Fourier expansion

$$f^\sigma(z) = \sum_{n=1}^{\infty} a_f(n)^\sigma e^{2\pi i n z}.$$

Following Momose [4] and Ribet [9], we consider the collection Γ_f of such σ for which there is a Dirichlet character $\chi = \chi_\sigma$ with the property that

$$a_f(p)^\sigma = a_f(p)\chi(p)$$

for almost all primes p. (We remark that as f is assumed to not be of CM-type, χ_σ is unique when it exists.) It is a fact that Γ_f is a finite Abelian group of automorphisms of E_f. We consider its fixed subfield F_f.

We may view each χ_σ as a character of $\mathrm{Gal}(\bar{\mathbb{Q}}/\mathbb{Q})$ with values in E_f^*. Let K_f denote the field fixed by all the kernels $\ker \chi_\sigma$ for $\sigma \in \Gamma_f$. Let $H_f = \mathrm{Gal}(\bar{\mathbb{Q}}/K_f)$ and denote by $H_{f,\ell}$ the image $\rho_{f,\ell}(H_f)$.

Then Ribet [11], Theorem 3.1, states that for ℓ sufficiently large, $H_{f,\ell}$ is equal to

$$A_{f,\ell} = \{x \in \mathrm{GL}_2(\mathcal{O}_{F_f} \otimes \mathbb{Z}_\ell) : \det x \in \mathbb{Z}_\ell^{*(k-1)}\}. \tag{3}$$

Let B_f be such that for $\ell > B_f$, the above statement is true.

5. Further results

Theorem 5.1. *With the hypotheses as in Theorem 3.1, and in particular, assuming the GRH, there exists a prime* p *with*

$$p \ll B_f^{14}(\log B_f N)^2(\log B_f)^2$$

for which $E_f = \mathbb{Q}(a_f(p))$.

The above result allows us to deduce the following corollary.

Corollary 5.1. *With* f *as above and assuming the GRH, suppose that for*

$$p \ll B_f^{14}(\log B_f N)^2(\log B_f)^2$$

we have $a_f(p) \in \mathbb{Z}$. *Then* $E_f = \mathbb{Q}$.

6. A geometric interpretation

In the case that the modular form has weight 2, there is a geometric interpretation of our results based on [8]. Indeed, let A_f be the Abelian variety associated to f. This is an Abelian variety defined (up to isogeny) over $\mathbb{Q}$ and with the property that

$$E_f \hookrightarrow \mathrm{End}_{\mathbb{Q}}(A_f) \otimes \mathbb{Q}$$

where the notation means that E_f acts $\mathbb{Q}$-rationally. Moreover,

$$[E_f : \mathbb{Q}] = \dim A_f.$$

It is a fact that A_f is $\mathbb{Q}$-simple and if N is squarefree and $\epsilon = 1$, it follows by the results of [10] and [4] that A_f is also $\bar{\mathbb{Q}}$-simple.

The connection of A_f with the modular form f comes through the Galois representation. This is most succinctly expressed in terms of the L-function. We have

$$L(A_f, s) = \prod_{\sigma \in \Sigma} L(f^\sigma, s).$$

Recall $\Sigma = \mathrm{Hom}(E_f, \mathbb{C})$. In other words, for primes p not dividing the level N, the p-Euler factor in the definition of the L-function is

$$\prod_{\sigma \in \Sigma} (1 - a_f(p)^\sigma p^{-s} + \epsilon(p) p^{1-2s})^{-1}.$$

Theorem 6.1. *With f as above of weight 2, and assuming the GRH, with the exception of $\mathbf{O}(x^{13/14})$ of the primes $p \leq x$, and for $p \nmid N$, the reduction $A_{f,p}$ of A_f modulo p is simple. Moreover, we can find a prime p for which $A_{f,p}$ is simple and*

$$p \ll B_f^{14}(\log B_f N)^2(\log B_f)^2.$$

It is possible to relax the hypotheses on f and obtain results about the generation of a certain subfield of E_f. We return to this at the end of the article.

7. The Chebotarev density theorem

The theorems are proved by applying the Chebotarev density theorem to various fields associated with the Galois representations described above. We begin this section with a general description of the Chebotarev density theorem.

Suppose we have a number field K which is Galois over $\mathbb{Q}$ with group G. Let C be a subset of G stable under conjugation. Thus, C is a union of conjugacy classes of G. Assuming the GRH, the Chebotarev density theorem in its effective form states that the number $\pi_C(x, K)$ of primes $p \leq x$ with the property that the Frobenius conjugacy class Frob_p is contained in C satisfies

$$\pi_C(x, K) = \frac{|C|}{|G|} \operatorname{Li} x + \mathbf{O}\left(\frac{|C|}{|G|} x^{\frac{1}{2}}(\log d_K + n_K \log x)\right)$$

where d_K denotes the absolute value of the discriminant of $K/\mathbb{Q}$ and $n_K = [K : \mathbb{Q}]$. The reader is referred to [6] for further results on the Chebotarev density theorem and for references.

To apply this in our context, we decompose the Galois representations associated to f as follows. Let us set $T_\ell = (\mathcal{O}_{E_f} \otimes \mathbb{Z}_\ell)^2$ and view $\rho_{f,\ell}$ as giving the action of $\mathrm{Gal}(\bar{\mathbb{Q}}/\mathbb{Q})$ on T_ℓ. Then,

$$T_\ell/\ell T_\ell = \prod_{\lambda|\ell} \mathbb{T}_\lambda$$

where $\mathbb{T}_\lambda = (\mathcal{O}_{E_f}/\lambda)^2$. We consider the composition of $\rho_{f,\ell}$ with reduction modulo λ to get

$$\overline{\rho_{f,\lambda}} : \mathrm{Gal}(\bar{\mathbb{Q}}/\mathbb{Q}) \longrightarrow \mathrm{GL}_2(\mathbb{F}_\lambda),$$

where we are writing $\mathbb{F}_\lambda = \mathcal{O}_{E_f}/\lambda$. From what we stated about $\rho_{f,\ell}$ at the beginning of section 4, it follows that $\overline{\rho_{f,\lambda}}$ is unramified outside of primes dividing ℓN. The above assertion (3) about $H_{f,\ell}$ implies that for ℓ sufficiently large, the image of $\overline{\rho_{f,\lambda}}$ is given by

$$\overline{A_\lambda} = \{x \in \mathrm{GL}_2(\mathbb{F}_\lambda) : \det x \in \mathbb{F}_\ell^{*(k-1)}\}.$$

Thus, the fixed field of the kernel of $\overline{\rho_{f,\lambda}}$, has Galois group isomorphic to $\overline{A_\lambda}$.

Moreover, choosing two primes λ_1 and λ_2 dividing ℓ, we consider the representation

$$\mathrm{Gal}(\bar{\mathbb{Q}}/\mathbb{Q}) \longrightarrow \mathrm{GL}_2(\mathbb{F}_{\lambda_1}) \oplus \mathrm{GL}_2(\mathbb{F}_{\lambda_2})$$

and denote by K the fixed field of the kernel of this map. We shall apply the Chebotarev density theorem to this field. The fact that the Galois representation is unramified outside primes dividing ℓN implies that

$$\log d_K \ll n_K \log \ell N.$$

We see that $\mathrm{Gal}(K/\mathbb{Q})$ is isomorphic to the image of $\mathrm{Gal}(\bar{\mathbb{Q}}/\mathbb{Q})$ on

$$\mathrm{GL}_2(\mathbb{F}_{\lambda_1}) \oplus \mathrm{GL}_2(\mathbb{F}_{\lambda_2})$$

and for ℓ sufficiently large, this image can be described explicitly as follows. If the restrictions $\lambda_1|_{F_f}$ and $\lambda_2|_{F_f}$ are not equal, then the image is given by

$$\overline{A_{\lambda_1,\lambda_2}} = \{(x_1, x_2) \in \mathrm{GL}_2(\mathbb{F}_{\lambda_1}) \times \mathrm{GL}_2(\mathbb{F}_{\lambda_2}) : \det x_1 = \det x_2 \in \mathbb{F}_\ell^{*(k-1)}\}.$$

On the other hand, if the restrictions are equal, then $\mathbb{T}_{\lambda_1}$ is isomorphic to $\mathbb{T}_{\lambda_2}$ and the image is the graph of an isomorphism.

Let us choose a prime ℓ that splits completely in E_f. In particular, all of the λ are of degree 1. In this case, the various λ are conjugate under the $\sigma \in \Sigma$. In this case, we see that

$$|\overline{A_{\lambda_1,\lambda_2}}| \sim \ell^7.$$

The notation means that there are absolute and effective constants so that

$$\ell^7 \ll |\overline{A_{\lambda_1,\lambda_2}}| \ll \ell^7.$$

Moreover, if we consider the subset C_{λ_1,λ_2} of elements (x_1, x_2) of $\overline{A_{\lambda_1,\lambda_2}}$ with the additional property that $tr\ x_1 = tr\ x_2$, then we see that C_{λ_1,λ_2} is a union of conjugacy classes and

$$|C_{\lambda_1,\lambda_2}| \sim \ell^6.$$

By the effective Chebotarev density theorem, as stated above, applied to this situation, and assuming the GRH, we see that

$$\#\{p \le x : tr\ \overline{\rho_{f,\lambda_1}}(\mathrm{Frob}_p) = tr\ \overline{\rho_{f,\lambda_2}}(\mathrm{Frob}_p)\} \ll \frac{1}{\ell}\ \mathrm{Li}\ x + \mathbf{O}(\ell^6 x^{\frac{1}{2}} \log \ell N x).$$

8. Proof of Theorem 3.1

Proof of Theorem 3.1. Let $d = [E_f : \mathbb{Q}]$ and consider the set $\Sigma = \mathrm{Hom}(E_f, \mathbb{C})$. It suffices to count the number of primes p for which the d conjugates $a_f(p)^\sigma, \sigma \in \Sigma$ are not distinct. For each pair $\sigma_1, \sigma_2 \in \Sigma$, consider the set

$$C_{\sigma_1,\sigma_2}(x) = \{p \le x : a_f(p)^{\sigma_1} = a_f(p)^{\sigma_2}\}.$$

Let ℓ be a prime that splits completely in E_f. Fix a prime λ of E_f dividing ℓ. Set

$$C_{\sigma_1,\sigma_2}(x, \lambda) = \{p \le x : a_f(p)^{\sigma_1} \equiv a_f(p)^{\sigma_2} \pmod{\lambda}\}.$$

We clearly have

$$|C_{\sigma_1,\sigma_2}(x)| \le |C_{\sigma_1,\sigma_2}(x,\lambda)|.$$

If we set $\lambda_1 = \sigma_1^{-1}\lambda$ and $\lambda_2 = \sigma_2^{-1}\lambda$, then we see that

$$|C_{\sigma_1,\sigma_2}(x,\lambda)| = \pi_{C_{\lambda_1,\lambda_2}}(x, K).$$

If

$$\sigma_1|_{F_f} \neq \sigma_2|_{F_f}, \tag{4}$$

then by the effective Chebotarev density theorem and assuming the Generalized Riemann Hypothesis, it follows from the discussion of the previous section that we have

$$|C_{\sigma_1,\sigma_2}(x,\lambda)| \ll \frac{1}{\ell}\,\mathrm{Li}\,x + \mathbf{O}(\ell^6 x^{\frac{1}{2}} \log \ell N x).$$

As this is uniform in ℓ, if we choose ℓ of the order $x^{1/14}/((\log Nx)(\log x))^{\frac{1}{2}}$, we deduce that

$$|C_{\sigma_1,\sigma_2}(x)| \ll x^{13/14}. \tag{5}$$

Under the hypotheses of the theorem (namely N is squarefree and $\epsilon = 1$), it follows by [10], (3.9), that $E_f = F_f$. Thus the condition (4) is satisfied by any pair of distinct embeddings of E_f. Thus, summing (5) over pairs (σ_1, σ_2) of distinct embeddings, we deduce that

$$\#\,\{p \le x : \ a_f(p) \text{ has } d \text{ distinct conjugates}\} = \mathrm{Li}\,x + \mathbf{O}(x^{13/14}).$$

This proves the result.

9. Proof of Theorem 5.1

We proceed as in the proof of Theorem 3.1. However, instead of choosing ℓ as a function of x, we choose x as a function of ℓ. Indeed, we take

$$x = \ell^{14}(\log \ell N)^2 (\log \ell)^2.$$

In this case, we find that for any pair of distinct embeddings σ_1, σ_2 of E_f, we have

$$|C_{\sigma_1,\sigma_2}(x)| \le |C_{\sigma_1,\sigma_2}(x,\lambda)| \ll \ell^{13}(\log \ell N)^2(\log \ell).$$

This is valid as long as $\ell > B_f$. Taking $\ell \in (B_f, 2B_f)$ (say), and summing over possible σ_1, σ_2, the result follows.

10. Proof of Theorem 6.1

Let A be an Abelian variety of dimension d defined over a number field K and denote by $\mathrm{End}_K(A)$ the ring of endomorphisms of A defined over K. Suppose that A has real multiplication in the sense that there is a number field E with $[E : \mathbb{Q}] = \dim A$ and

$$E \hookrightarrow \mathrm{End}_K(A) \otimes \mathbb{Q}.$$

Assume also that under this map, $1 \in E$ is sent to the identity element. Then, for a prime v of K where A has good reduction and for which the underlying rational prime $p = p(v)$ does not ramify in E, we have the following result (see [8], Proposition 2.2).

Proposition 10.1. *The characteristic polynomial of Frobenius associated to the reduction A_v of A modulo v has repeated roots if and only if A_v is not simple over the residue field k_v of K at v.*

Applying this to the Abelian variety A_f, we first observe that the hypotheses are satisfied with $K = \mathbb{Q}$ and $E = E_f$. Moreover, for any prime $p \nmid N$, we know that A_f has good reduction and the characteristic polynomial of Frobenius acting on $A_{f,p}$ is

$$\prod_{\sigma}(T^2 - a_f(p)^{\sigma} T + p).$$

It is easy to deduce (see [8], Proof of Theorem 4.1, p. 21) that for this polynomial to have repeated roots means that for some $\sigma_1 \neq \sigma_2$, we have

$$a_f(p)^{\sigma_1} = a_f(p)^{\sigma_2}.$$

In ([8], Theorem 4.3), it is shown that this condition implies that $\Gamma_f \neq \{1\}$. By a result of Ribet ([10], equation (3.9)), this cannot happen under the hypotheses of Theorem 6.1. This proves the result.

11. Concluding remarks

We conclude with some remarks about refinements and variants of the above results.

The hypotheses N squarefree and $\epsilon = 1$ were only made to ensure that $\Gamma_f = \{1\}$. In general, recall that we have set K_f to be the fixed field of the kernels of all the twisting characters χ_σ (for $\sigma \in \Gamma_f$). The same methods as used above allow us to show that amongst the primes p that split completely in the field K_f, a set of density one has the property that $a_f(p)$ generates F_f.

The effective estimates of Theorem 3.1 can be improved. Indeed, using the methods of [6], we expect that the result of Theorem 3.1 can be improved to $\mathbf{O}(x^{8/9})$ (assuming the GRH). Using the results of [7], the unconditional estimate can be improved as well.

The case of half-integral weight modular forms was briefly alluded to in section 2. Under some hypotheses on the cusp form, it is shown in [3] that the Fourier coefficients of a half-integral weight form (suitably normalized) lie in a number field. One may ask whether a single Fourier coefficient generates that field. A related problem has to do with integral weight forms. Let f be a normalized eigenform of weight k and χ a Dirichlet character. Denote by $L(f, \chi, s)$ the L-function associated to f twisted by χ. A result of Shimura states that the special value $L(f, \chi, k/2)/\Omega_{\pm}$ is algebraic and takes values in $E_f(\chi)$ (the field generated over E_f by values of χ). Here, $\Omega_{\pm}$ are periods associated to f and we choose the one corresponding to the sign of the character χ. One may ask whether this special value generates this field.

In [5], another variant is studied, namely for congruences between modular forms. In other words, given a prime v, and two forms f and g that are not congruent modulo v, one may ask for an estimate of the least n for which

$$a_f(n) \not\equiv a_g(n) \pmod{v}.$$

The analogous question here is about the estimation of the least n for which the $a_f(n)^\sigma$ are distinct modulo a prime v.

References

[1] K. T.-L. Koo, W. Stein and G. Wiese, On the generation of the coefficient field of a newform by a single Hecke eigenvalue, *J. Théor. Nombres Bordeaux*, **20(2)** (2008) 373–384.

[2] M. Manickam, V. Kumar Murty and B. Ramakrishnan, Non-vanishing of L-functions of half-integral weight, preprint (2011).

[3] M. Manickam, V. Kumar Murty and B. Ramakrsihnan, New inner products on spaces of modular forms, preprint (2011).

[4] F. Momose, On the ℓ-adic representations attached to modular forms, *J. Fac. Sci. Univ. Tokyo Sect. 1 A Math.*, **28** (1981) 89–109.

[5] M. Ram Murty, Congruences between modular forms, In: *Analytic Number Theory*, pp. 309–320, ed. Y. Motohashi, London Mathematical Society Lecture Notes **247**, Cambridge University Press, Cambridge (1997).

[6] M. Ram Murty, V. Kumar Murty and N. Saradha, Modular forms and the Chebotarev density theorem, *Amer. J. Math.*, **110** (1988) 253–281.

[7] V. Kumar Murty, Modular forms and the Chebotarev density theorem, II, In: *Analytic Number Theory*, 287–308, ed. Y. Motohashi, *London Math. Soc. Lecture Notes*, **247**, Cambridge University Press, Cambridge (1997).

[8] V. Kumar Murty and V. Patankar, Splitting of Abelian varieties, *Intl. Math. Res. Notices* **12**, Article ID rnn033, 27 pages (2008).

[9] K. Ribet, Galois representations attached to modular forms, *Invent. Math.*, **28** (1975) 245–275.

[10] K. Ribet, Twists of modular forms and endomorphisms of Abelian varieties, *Math. Annalen*, **253** (1980) 43–62.

[11] K. Ribet, On ℓ-adic representations attached to modular forms II, *Glasgow Math. J.*, **27** (1985) 185–194.

TCV Proceedings No. 15, 2011, pp. 113–117.

Indefinite Universal Quadratic Forms

S. Raghavan

Flat S1, Suryakantha, 26, C.I.T. Colony, Second Main Road, Mylapore,
Chennai 600 004, India
e-mail: navgraha@bsnl.in

As a surprising sequel to Lagrange's famous four squares theorem [5] that every natural number is a sum of 4 squares of integers (equivalently, the quaternary definite quadratic form $X^2 + Y^2 + Z^2 + U^2$ represents over $\mathbf{Z}$ all the natural numbers or is briefly 1-*universal* or just *universal*, Ramanujan's important paper captured all the (54 '$\mathbf{Z}$-equivalence classes' of) integral diagonal definite 1-universal quaternary quadratic forms including indeed the form just mentioned. Later M. F. Willerding [9] seeking to extend Ramanujan's result, found in all 178 (not necessarily diagonal) quaternary universal forms of the same kind, while at the same time, missing out on 36 further such forms, as we learn now from Bhargava's remarkable paper [1].

As an analogue of Lagrange's celebrated theorem, Mordell's "five square theorem" [6] in 1930 showed that the diagonal quinary form $X^2 + Y^2 + Z^2 + U^2 + V^2$ is 2-*universal*, i.e. all 'classical' positive definite binary quadratic forms $aP^2 + 2bPQ + cQ^2$ (i.e. with integral a, b, c) are sums of five squares of linear forms in X, Y, Z, U and V with integral coefficients. For further recent developments concerning "k-*universal*" positive definite integral quadratic forms for $k \geq 1$, one may read the excellent article [4].

In connection with [2], Prof. Basil Gordon directed our attention to the case of indefinite integral quadratic forms which are 'universal' (i.e. representing integrally all non-zero integers) and raised the question as to the existence of 'universal' binary indefinite forms 'apart from' the simplest 'universal' (but *not* classical) binary quadratic form XY (A quadratic form is called *classical integral* or just *classical* if its associated (2,2) symmetric matrix has all its entries in $\mathbf{Z}$). This short write-up seeks, *inter alia*, to address Prof. Gordon's interesting question.

A non-degenerate quadratic form f in n variables $X_1, \ldots, X_n$ with real coefficients $a_{ij} = a_{ji}$ has *signature* (p, q) in the sense of Siegel [8], if p (respectively q) of its eigenvalues are positive (respectively negative) and is *indefinite* if $pq > 0$. Such an f when indefinite and all its coefficients a_{ij} are integral is said to be *universal*, if the value set $\mathscr{W}(f) := \{f(a_1, \ldots, a_n)$ with $a_1, \ldots, a_n$ independently running over $\mathbf{Z}\}$ contains $\mathbf{Z}\backslash\{0\}$. More generally, we call such an f k-*universal* (for $k = 1$ or > 1), if for every classical non-zero quadratic form $g = g(Y_1, \ldots Y_k)$, we can express $X_1, \ldots, X_n$ as suitable linear forms in $Y_1, \ldots Y_k$ with all coefficients in $\mathbf{Z}$, such that $f(X_1, \ldots, X_n) = g(Y_1, \ldots Y_k)$. Whenever $f(X_1, \ldots, X_n) = g(Y_1, \ldots Y_k)$ in this manner, we say that f *represents* g. If f represents g and *vice versa*, we call f and g

Z-*equivalent* and such an equivalence is called an *isometry* (which necessarily entails $k = n$). Indefinite forms considered here are non-degenerate. By an *even* quadratic form in n variables $X_1, \ldots, X_n$, we mean a classical integral quadratic form for which the coefficients of X_i^2 are all even integers for $i = 1, \ldots, n$. Let $u(n)$ (respectively $u_2(n)$) be the least number m of variables such that there exists a classical integral indefinite quadratic form in m variables which represents all classical integral (respectively even) quadratic forms in n variables. It is an interesting question to determine $u(n)$ and $u_2(n)$. From Theorem 2, we will see that $u_2(n) = 2n$. As an immediate corollary of Theorem 1, we will see that $u(1) > 2$. However to conclude that $u(1) = 3$, we quickly note first that for $f(X, Y) := X^2 - Y^2$, we have $\mathscr{W}(f) = \{$all integers r not of the form $2u$ with u odd$\}$, as checked easily and so f is not universal. However, it follows immediately that the indefinite ternary form $X^2 - Y^2 - Z^2$ is indeed universal and likewise so is any other form $X^2 - Y^2 - mZ^2$, provided that m is an integer not divisible by 4. (One may incidentally be tempted to jump to the conclusion that in contrast with the case of definite forms, the relevant Conway–Schneeberger–Bhargava set may have to be infinite. We hope to come back to this question later).

It was a message from Prof. M.-H. Kim that sounded the alert on the existence of such an infinity of isometry classes of universal indefinite ternary forms in sharp contrast with the case of definite universal forms (Prop. 2 of [3]). He also stated "$u(2) = 5, u(n) = 2n$ for $n > 2$". No classical indefinite quaternary quadratic form can be 2-universal as may be checked (e.g. by splitting off a binary form $2XY$) and so, by Theorem 2 below, we have $u(2) = 5$. Prof. W.-K. Chan has recently pointed out that the fact "*XY* is the only integral binary universal form"(also, over any number field) was proved by the late Dennis Estes and his student Albert Lee in 1993 (Classifying rank 2 quadratic lattices by representations, Nova Journal of Algebra and Geometry, 2 (1993), 201–217). Further, Prof. Chan wrote: "The indefinite ternary case was handled much earlier by Dickson in Theorem 109 of his book "Modern Elementary Theory of Numbers". He showed that up to equivalence, the indefinite integral ternary universal forms are $X^2 - Y^2 - aZ^2$ with $\text{ord}_2(a) = 0$ or 1, and $XY - bZ^2$".

We may now recall that for a binary form $f(X, Y) := aX^2 + bXY + cY^2$ with its coefficients a, b, c integral, its discriminant $d(f) := (b^2 - 4ac)/4$ equals $t^2/4$ for integral t if and only if f is a so-called zero form, i.e. f represents 0 non-trivially over **Z**; in that case, passing to an equivalent form, we may, without loss of generality, write f as $rXY + sY^2$ with integral r, s. We seek the readers' indulgence for including Theorem 1 below, despite the availability of the result even in greater generality thanks to Estes and Lee.

Theorem 1. *Every universal indefinite integral quadratic form* $f(X, Y) := aX^2 + bXY + cY^2$ *is isometric to* XY.

Proof. Since $1 \in \mathscr{W}(f) = Z\backslash\{0\}$, we may ,by passing on to an isometric form if necessary, assume already that $a = 1$. Clearly, under any isometry, the parity of the middle coefficient b is preserved and so is $d := d(f)$.

We claim now that b must be odd due to f being universal. Let, if possible, b be even so that f becomes $(X + (b/2)Y)^2 - dY^2$. Then we can indeed pass to the isometric $X^2 - dY^2$ with, of course, $d > 0$. First, if $d = t^2$ for integral $t > 0$, then the remark on $\mathscr{W}(f)$ immediately preceding Theorem 1 rules out the case $t = 1$; if, however, $t > 1$, then f cannot be universal, since no prime divisor p of t can belong to $\mathscr{W}(f)$. Next, if $d > 1$ is not the square of an integer, there exists an odd prime p such that d is a quadratic non-residue modulo p, contradicting our requirement that p too has to belong to $\mathscr{W}(f)$. Thus our claim about b being odd is verified and we may start with $f(X, Y) := X^2 + bXY + cY^2$ universal and with odd b. Now $4d = b^2 - 4c$ is congruent to 1 modulo 4. If $4d = t^2$ for integral $t > 0$, then clearly $t - b$ is even and therefore f is isometric to $X^2 + tX\,Y$. For $t = 1$, we are through, since f is then isometric to XY. If $t > 1$, we take a prime factor q of t and have, from f being universal, integers k, l such that $k^2 + tkl = q$ and so q^2 divides q! The contradiction proves the Theorem if $4d$ is a square. On the other hand, if $4d$ is not the square of an integer, we can find an odd prime q for which $4d$ is a quadratic non-residue. But f being universal, we have integers k, l such that $(2k + bl)^2 - 4d\,l^2 = 4q$, leading to a contradiction and finally proving Theorem 1.

The following theorem includes an analogue of Mordell's "five square theorem" [6].

Theorem 2. *The indefinite quadratic form $X_1Y_1 + X_2Y_2 + \cdots + X_nY_n$ represents integrally every integral (not necessarily classical) quadratic form $g(P_1, P_2, \ldots, P_n)$ in n variables. Thus $u_2(n) = 2n \le u(n) \le 2n + 1$.*

Proof. If $g(P_1, P_2, \ldots, P_n) = \sum_{1\le i\le n}(a_{ii}P_i^2 + a_{ii+1}P_iP_{i+1} + \cdots + a_{in}P_iP_n)$, we simply need to take $X_i = P_i$, $Y_i = \sum_{i\le j\le n} a_{ij}P_iP_j$. In other words, every integral quadratic form in n variables can be written as a sum of n products of 2 integral linear forms and thus representable integrally by the given indefinite quadratic form of signature (n, n) in $2n$ variables. Observe now that for any integral quadratic form of signature (p, q) to represent all (classical) integral quadratic forms in n variables and, in particular, those of signature $(n, 0)$ or $(0, n)$, $2n$ is the minimum value for the number $p+q$ of variables and so $2n \le u(n)$. But since $2X_1Y_1 + 2X_2Y_2 + \cdots + 2X_nY_n$ clearly represents all even quadratic forms, it follows that $u_2(n) = 2n$. To see that $u(n) \le 2n + 1$, we take any form g as written above but classical and with notation changed; namely, without loss of generality, if, for a suitable integer m between 0 and n, its first m diagonal coefficients a_{ii} are even and the last $n - m$ diagonal coefficients are odd, we rewrite them as $2b_{11}, 2b_{22}, \ldots, 2b_{mm}, 2b_{m+1,m+1} + 1, \ldots, 2b_{nn} + 1$; we also rewrite the (necessarily even) coefficients of P_rP_s for $r < s$ as $2(b_{rs} + 1)$ or $2b_{rs}$ according as $r \le m < s$ or not (with the integers b_{rs} for $r < s$ defined thus in the process). It is easy to check then that g takes the form

$$
\begin{aligned}
&g(P_1, P_2, \ldots, P_n)\\
&\quad = \sum_{1\le r\le m}\left(2b_{rr}P_r^2 + 2\sum_{r<s\le m} b_{rs}P_rP_s + 2\sum_{m<s\le n}(b_{rs} + 1)P_rP_s\right)\\
&\qquad + \sum_{m<r\le n}\left((2b_{rr} + 1)P_r^2 + 2\sum_{r<s\le n} b_{rs}P_rP_s\right)
\end{aligned}
$$

$$= \sum_{1\le j\le m}\left[\left\{\sum_{j\le k\le n}(b_{jk}+1)P_k\right\}^2 - \left\{b_{jj}P_j + \sum_{j<k\le n}(b_{jk}+1)P_k\right\}^2\right]$$

$$-(P_1+\cdots+P_m)^2 + \sum_{m+1\le j\le n}\left[\left\{P_j + \sum_{j\le k\le n} b_{jk}P_k\right\}^2 - \left\{\sum_{j\le k\le n} b_{jk}P_k\right\}^2\right]. \qquad \text{(MT)}$$

From this identity which we refer to as (MT), we conclude that the indefinite integral quadratic form (in $2n+1$ variables) $f(X_1, X_2, \ldots, X_n, Y_1, Y_2, \ldots, Y_n, Z) := \sum_{1\le i\le m}(X_i^2 - Y_i^2) - Z^2 + \sum_{m+1\le j\le n}(X_j^2 - Y_j^2)$ represents g integrally leading us to $u(n) \le 2n+1$. Regrettably, identity (MT) can yield no more, say, confirmation that (possibly) $u(n) = 2n$ for $n > 2$! However, Prof. B. K. Oh pointed out recently that the required confirmation would follow at once from the n-universality, for $n > 2$, of $I(n,n) := \sum_{1\le i\le n}(X_i^2 - Y_i^2)$ by taking $f = I(n,n)$ in the result that for integral quadratic forms g of rank n and f of rank $m > n+2$, everywhere local integral representability of g by f implies the global representability of g by the spinor genus of f (which contains just one class in view of f being indefinite).

Remarks. We briefly review the scenario with $\mathbf{Z}$ replaced by the ring R of integers in an algebraic number field K of degree t over $\mathbf{Q}$. In view of the results of D. Estes and A. Lee cited above, we see that for $u(n)$ and $u_2(n)$ defined over R analogously, the claims "$u_2(n) = 2n \le u(n)$" go through in exactly the same way.

However, with a suitably modified form of identity (MT) above, we can prove that $u(n) \le 2n + 1 + \lambda(n,t)$ where $\lambda = \lambda(n,t) := (2^t - 2)^n$. Indeed, there exists a suitable integer m between 0 and n such that the first m diagonal coefficients of g can be assumed to be $2b_{11}, 2b_{22}, \ldots, 2b_{mm}$ (belonging to $2R$) while the next $n-m$ diagonal coefficients are $2b_{m+1,m+1} + c_{m+1}, \ldots, 2b_{nn} + c_n$ where $c_{m+1}, \ldots, c_n$ lie in $R\backslash 2R$ and can be taken to be in a fixed representative system $\mathscr{R}$ including 1 (of cardinality $2^t - 1$) for R modulo $2R$. Now $g(P_1, P_2, \ldots, P_n)$ takes the same form as before except that the diagonal coefficients are $2b_{rr} + c_r$ instead of $2b_{rr} + 1$ for $m < r \le n$. On the right hand side of the modified identity (MT), there occurs now an additional summand $\sum_{m<r\le n}(c_r - 1)P_r^2$. Since m can equal n at the most, we have no more than $(2^t - 2)^n$ possibilities for such (non-zero) additional terms. Clearly $\sum_{1\le i\le m}(X_i^2 - Y_i^2) - Z^2 + \sum_{m+1\le j\le n}(X_j^2 - Y_j^2) + d_1 Z_1^2 + \cdots + d_\lambda Z_\lambda^2$ with d_i from $\mathscr{R}\backslash\{1\}$ represents $g(P_1, P_2, \ldots, P_n)$ over R proving the assertion "$u(n) \le 2n + 1 + \lambda(n,t)$". For $t > 1$, we end up only with a poor estimate; although $\lambda(n,1) = 0$, $\lambda(n,t)$ is exponential in n and t. Fortunately however, the n-universality of the diagonal matrix $I(n,n) = \langle 1, 1, \ldots, 1, -1, -1, \ldots, -1\rangle$ of signature (n,n) for $n > 2$ happens to hold over number fields! In his famous Annals Math. 45 (1944) paper, Siegel showed that for integral symmetric (n,n) matrices T of rank n and (m,m) *indefinite* S of signature (p,q) and rank $m = p + q (> 2n+2)$ with $p > n-1$ and $q > n-1$, "the measure $M(S,T)$ of representation of T by S" *depends only on the genus of* S. From this theorem of Siegel's, one can see that $u(n) < 2n+4$. Identity (MT) above yields the slightly stronger result "$u(n) < 2n + 2$" (but not the 'ultimate' $u(n) = 2n$ for $n > 2$)! Weil's paper in (Colloque sur la theorie des groupes algebriques, Bruxelles, 1962), with remarkable adelic and algebraic group-theoretic methods, went further than Siegel's formulae in the analytic theory of quadratic forms (including the one for indefinite forms in his Annals 1944 paper) by extending their *validity* to $m > n + 2$ and also to *averages over classes in a spinor genus* (*instead of genus*) and further dispensing with the *conditions* "$p > n - 1$ and $q > n - 1$" mentioned above. Considering (only non-degenerate) integral quadratic forms now over algebraic number fields, Kneser's result that indefinite spinor genera (for 3 or more variables) contain only one class taken together with the above result of Weil's (over number fields) as applied to $S = I(n, n)$ with $n > 2$ and any T of rank n (and the fact that no local density of representation of T by $I(n, n)$ vanishes while the relevant infinite product of p-adic densities converges (to a positive number)), lead us to the non-vanishing of $M(S, T)$ for the spinor genus (i.e. the class) of $S = I(n, n)$ with $n > 2$ for all T above and finally $u(n) = 2n$ for $n > 2$ valid over number fields too.

References

[1] M. Bhargava, On the Conway-Schneeberger Fifteen Theorem, *A. M. S. Contemp. Math.*, **272** (2000) 27–38.

[2] B. M. Kim, M.-H. Kim and S. Raghavan, 2-universal positive definite integral quinary diagonal quadratic forms, *Ramanujan J.*, **1** (1997) 333–337.

[3] B. M. Kim, M.-H. Kim and B.-K. Oh, 2-universal positive definite integral quinary quadratic forms, *A. M. S. Contemp. Math.*, **249** (1999) 51–62.

[4] M.-H. Kim, Recent developments on universal forms, *A. M. S. Contemp. Math.*, **344** (2002) 215–228.

[5] J. L. Lagrange, Oeuvres, *Tome*, **3** (1869).

[6] L. J. Mordell, A new Warings problem with squares of linear forms, *Quart. J. Math. Oxford*, **1** (1930) 276–288.

[7] S. Ramanujan, On the expression of a number in the form $ax^2 + by^2 + cz^2 + du^2$, *Proc. Camb. Phil. Soc.*, **19** (1917).

[8] C. L. Siegel, Zur Theorie der quadratischen Formen, Gesamm. Abhand. IV, Springer Verlag (1979) 224–249.

[9] M. F. Willerding, Determination of all classes of positive quaternary quadratic forms which represent all (positive) integers, *Bull. Amer. Math. Soc.*, **54** (1948) 334–337.

TCV Proceedings No. 15, 2011, pp. 119–141.

Companion Forms and Classicality in the $\mathbf{GL}_2(\mathbb{Q})$-Case

J. Tilouine*

Département de Mathématiques, UMR 7539, Institut Galilée, Université de Paris 13, 93430 Villetaneuse, France
e-mail: tilouine@math.univ-paris13.fr

Abstract. We give a new proof of the existence of a companion form for a modular form of weight $k \in [2, p[$ whose residual Galois representation is semisimple at p, by using the BGG dual complex. We apply this to the case of weight 1 p-adic over convergent forms by using deformation theory.

Keywords. mod p modular forms, p-adic modular forms, Galois representations

To Prof. T. C. Vasudevan for his Sixtieth Birthday

1. Introduction

In this paper we give proofs of the existence of companion forms in the elliptic modular case. We study two situations. The first concerns modulo p classical cusp eigenforms in weight $k \in [2, p-1]$; there, our proof is based on the modulo p étale-crystalline comparison theorem, in the spirit of the approaches of [14,11] and [4] (instead of the original one [17]). The second concerns p-adic overconvergent forms in weight $k = 1$. In this case the proof is based on a deformation theory argument, as in the paper of Buzzard-Taylor [7] and its generalization [8], both devoted to the study of Artin's conjecture in degree 2. Our approach of the first part differs from the previous ones by its use of integral and rigid structures of the dual BGG complex which clarify in our opinion some calculations of [11] and [4]. Let us recall that we already used this tool in an essential way in another situation, in [27]. This work is a translation in the elliptic modular case of a work in progress in the genus two Siegel modular case, where we establish similar results in the course of the study of Yoshida's conjecture ([34] and [32]).

This paper was written in part during a visit at RIMS in Kyōto and at Hokkaido University, where the excellent conditions of work were appreciated.

2. Modular forms with p-small weight $k \geq 2$ and cohomology

Let $N \geq 1$ and $\Gamma_1(N) = \{\gamma = \left(\begin{smallmatrix} a & b \\ c & d \end{smallmatrix}\right) \in SL_2(\mathbb{Z}); d \equiv 1, C \equiv 0 \pmod{N}\}$.

Let f be a fixed cusp eigenform of level group $\Gamma_1(N)$ with $N \geq 1$ prime to p, of weight $k \geq 2$ and character $\epsilon : (\mathbb{Z}/N\mathbb{Z})^\times \to \mathbb{C}^\times$. We shall write in the sequel

*Partially supported by Grant ANR-10-BLAN-0114.

$k = n+2$ with $n \geq 0$. Let p be a prime which does not divide N and such that $k < p$ that is, $k-1 = n+1 < p-1$. We fix an algebraic closure $\bar{\mathbb{Q}}_p$ of $\mathbb{Q}_p$; we endow it with the p-adic valuation ord_p such that $\mathrm{ord}_p(p) = 1$. We fix an embedding ι_p of the field $\bar{\mathbb{Q}}$ of algebraic complex numbers into $\bar{\mathbb{Q}}_p$. We assume that f is ordinary with respect to this embedding; this means that $f|T_p = a_p \cdot f$ and $\mathrm{ord}_p(\iota_p(a_p)) = 0$. Let us put $a_p = \alpha + \beta$ with $\alpha\beta = p^{k-1}\epsilon(p)$ and $\mathrm{ord}_p(\iota_p(\alpha)) = 0$.

Let $\rho_f : \mathrm{Gal}(\bar{\mathbb{Q}}/\mathbb{Q}) \to {}_2(\bar{\mathbb{Q}}_p)$ be the p-adic Galois representation associated to (f, p). Let us fix a p-adic discrete valuation subring $\mathcal{O} \subset \bar{\mathbb{Q}}_p$ containing the Hecke eigenvalues of f and over which the representation ρ_f is defined. Let $\chi : \mathrm{Gal}(\bar{\mathbb{Q}}/\mathbb{Q}) \to \mathbb{Z}_p^\times$ resp. $\omega : \mathrm{Gal}(\bar{\mathbb{Q}}/\mathbb{Q}) \to \mathbb{F}_p^\times$, be the p-adic, resp. modulo p cyclotomic character.

Recall that P. Deligne in a letter to Serre (May 28th, 1974) proved that the assumption of ordinarity of f at ι_p implies that the restriction of ρ_f to a decomposition group D_p at p has a special form. For any $\gamma \in \mathcal{O}^\times$, we define $\xi(\gamma) : D_p \to \mathcal{O}^\times$ to be the unramified character sending a geometric Frobenius at p to γ. Then,

$$\rho_f|D_p \sim \begin{pmatrix} \xi(\alpha) & * \\ 0 & \xi\left(\frac{\beta}{p^{k-1}}\right)\chi^{1-k} \end{pmatrix}$$

Let ϖ be a uniformizing parameter of $\mathcal{O}$ and $\mathbb{F} = \mathcal{O}/(\varpi)$ its residue field. We assume throughout this paper that the residual representation $\bar{\rho}_f : \mathrm{Gal}(\bar{\mathbb{Q}}/\mathbb{Q}) \to \mathrm{GL}_2(\mathbb{F})$ is irreducible. The first part of the paper (Sections 2, 3, 4) is devoted to the proof of the following Theorem, due to Gross [17].

Theorem 1. *If $\bar{\rho}_f$ is tamely ramified, and if $k < p-1$, then there exists an ordinary cusp eigenform g of level $\Gamma_1(N)$ and weight $k' = p+1-k$ with Hecke eigenvalues b_n, such that for any $n \geq 1$, prime to pN, we have $a_n \equiv n^{k-1} b_n (\mathrm{mod}\ \mathfrak{p})$, where $\mathfrak{p}$ is the prime of $\bar{\mathbb{Z}}$ determined by ι_p.*

If $k = p-1$, the same conclusion holds except that the ordinary cusp form g of weight $k' = 2$ may have level N or N_p, in which case we can replace it by a form of weight $p+1$ and level N.

This can be reformulated as a confirmation of the (now proved) Serre's Residual Modularity Conjecture as follows. Indeed, by Deligne's formula above, we have

$$\bar{\rho}_f|D_p \sim \begin{pmatrix} \xi(\bar{\alpha}) & * \\ 0 & \xi\left(\frac{\beta}{p^{k-1}}\right)\omega^{1-k} \end{pmatrix}$$

with the upper right shoulder being zero; consider the Galois representation $\bar{\rho} = \bar{\rho}_f \otimes \omega^{k-1}$; it is odd, continuous, irreducible and its restriction to the inertia subgroup I_p is conjugated to

$$\bar{\rho}|I_p \sim \begin{pmatrix} 1 & * \\ 0 & \omega^{1-k'} \end{pmatrix}$$

(with $* = 0$ and $k' = p+1-k \in [2, p-1]$), so its Fontaine-Serre's weights are 0 and $k'-1$; it must therefore come from a ι_p-ordinary cusp eigenform g of level N

prime to p, and weight $k' < p$:

$$\bar{\rho}_f \otimes \omega^{k-1} = \bar{\rho}_g.$$

Comparing the traces on Frobenius elements at $\ell \nmid N_p$, and using Brauer-Nesbitt theorem, we see that this statement is equivalent to Gross theorem as stated above: $a_n \equiv b_n n^{k-1} \pmod{\mathfrak{p}}$ for all n's prime to N_p.

2.1 *Integral modular forms*

We briefly recall some classical definitions. More details can be found in [24,18,19] and [20]. Let X be the modular curve defined over $\mathbb{Z}[\frac{1}{N}]$ which classifies elliptic curves endowed with a point of order exactly N. If $N > 4$, it is a fine moduli scheme. We assume this for the moment. Then, it is a geometrically connected smooth quasi-projective scheme endowed with a universal elliptic curve $f : \boldsymbol{E} \to X$. Let $\bar{X}$ be the arithmetic smooth compactification over $\mathbb{Z}[\frac{1}{N}]$ of X obtained by adding the relative divisor $C = \mathbf{Cusps}$ of cusps (see [24]); the divisor C is finite at over $\mathbb{Z}[\frac{1}{N}]$. This compactification is endowed with a universal one dimensional semi-abelian group scheme (the neutral connected component of the universal generalized elliptic curve) $f : \mathcal{G} \to \bar{X}$ with a section of order exactly N. Let e be the zero section of f. The sheaf of relative differentials $\omega = e^*\Omega_{\mathcal{G}/\bar{X}}$ is locally free of rank one. Let $\pi : \mathcal{T}_\omega \to \bar{X}$ be the $\mathbb{G}_m$-torsor given by $\mathcal{T}_\omega = \mathrm{Isom}_{\bar{X}}(\mathcal{O}_{\bar{X}}, \omega)$. The $\mathcal{O}_{\bar{X}}$-Module $\pi_*\mathcal{O}_{\mathcal{T}\omega}$ is a representation of the group scheme $\mathbb{G}_m$; any such representation is totally decomposed by the characters of the torus $X^*(\mathbb{G}_m) = \mathbb{Z}$; for any $k \in \mathbb{Z}$, let $\chi_k : t \mapsto t^k$ be the corresponding character of $\mathbb{G}_m$. Then we have $\pi_*\mathcal{O}_{\mathcal{T}\omega} = \oplus_{k\in\mathbb{Z}} \pi_*\mathcal{O}_{\mathcal{T}\omega}[\chi - k]$. Let us put $\omega^k = \pi_*\mathcal{O}_{\mathcal{T}\omega}[\chi - k]$; it is an invertible sheaf; we define the subsheaf $\omega_k = \omega^k(-\mathbf{Cusps})$. For any $\mathbb{Z}[\frac{1}{N}]$-algebra R, we define the modules of arithmetic modular resp. cuspidal forms by

$$M_k(\Gamma, R) = H^0(\bar{X} \times R, \omega^k) \quad \text{and} \quad S_k(\Gamma, R) = H^0(\bar{X} \times R, \omega_k).$$

2.2 *Hodge filtration and Gauss-Manin connection*

We recall definitions by stressing the fact that they are valid over the ring $\mathbb{Z}[\frac{1}{N}]$, hence over any $\mathbb{Z}[\frac{1}{N}]$-algebra.

Let $\bar{f} : \bar{\boldsymbol{E}} \to \bar{X}$ be the Kuga-Sato compactification of $\boldsymbol{E} \to X$; the relative surface $\bar{\boldsymbol{E}}$ is regular and the smooth locus of $\bar{f}$ is $\mathcal{G} \subset \bar{\boldsymbol{E}}$; moreover, the inverse image $D = \bar{f}^{-1}(C) = \bar{\boldsymbol{E}}\backslash\boldsymbol{E}$ is a divisor with normal crossings in $\bar{\boldsymbol{E}}$. Let $\mathcal{H} = R^1\bar{f}_*\Omega^\bullet_{\bar{\boldsymbol{E}}/\bar{X}}(d\log D/C)$. It is a rank two locally free $\mathcal{O}_{\bar{X}}$-sheaf together with a two-step filtration -the Hodge filtration, and a log-connection -the Gauss-Manin connection, $\nabla : \mathcal{H} \to \mathcal{H} \otimes \Omega_{\bar{X}}(d\log C)$. Let us recall the definition of this filtration and this log-connection.

First, the Hodge filtration is simply given by the short exact sequence

$$0 \to \bar{f}_*\Omega^1_{\bar{\boldsymbol{E}}/\bar{X}}(d\log D/C) \to R^1\bar{f}_*\Omega^\bullet_{\bar{\boldsymbol{E}}/\bar{X}}(d\log D/C) \to R^1\bar{f}_*\mathcal{O}_{\bar{\boldsymbol{E}}} \to 0$$

that is, $\mathrm{Fil}^i = \mathcal{H}$ if $i \leq 0$, $\mathrm{Fil}^1 = \bar{f}_*\Omega^1_{\bar{E}/\bar{X}}(d\log D/C)$, and $\mathrm{Fil}^i = 0$ if $i > 1$. Note that a trivial calculation in local charts at cusps shows that $\bar{f}_*\Omega^1_{\bar{E}/\bar{X}}(d\log D/C) = e^*\Omega_{\mathcal{G}/\bar{X}} = \omega$. Then by Serre duality, we have $R^1\bar{f}_*\mathcal{O}_{\bar{E}} \cong \omega^{-1}$, hence, $gr^0\mathcal{H} = \omega^{-1}$ and $gr^1\mathcal{H} = \omega$.

The Gauss-Manin connection is defined as follows. Consider the short exact sequence

$$0 \to \bar{f}^*\Omega^1_{\bar{X}}(d\log C) \to \Omega^1_{\bar{E}}(d\log D) \to \Omega^1_{\bar{E}/\bar{X}}(d\log D/C) \to 0$$

It gives rise to a filtration of the log-de Rham complex $\Omega^\bullet_{\bar{E}}(d\log D)$ by $\mathrm{Fil}^i = \mathrm{Im}(\bar{f}^*\Omega^i_{\bar{X}}(d\log C) \otimes \Omega^{\bullet-i}_{\bar{E}}(d\log D) \to \Omega^\bullet_{\bar{E}}(d\log D))$ for $i = 0, 1, 2$.

The associated graded pieces are

$$gr^i = \bar{f}^*\Omega^i_{\bar{X}}(d\log C) \otimes \Omega^{\bullet-i}_{\bar{E}/\bar{X}}(d\log D/C)$$

$(i = 0, 1)$.

In particular we have a short exact sequence of complexes

$$0 \to \bar{f}^*\Omega^1_{\bar{X}}(d\log C) \otimes \Omega^{\bullet-1}_{\bar{E}/\bar{X}}(d\log D/C)$$
$$\to \Omega^\bullet_{\bar{E}}(d\log D) \to \Omega^\bullet_{\bar{E}/\bar{X}}(d\log D/C) \to 0$$

The log-connection ∇ is defined as the connecting morphism associated to $R\bar{f}_*$ applied to this short exact sequence; it reads:

$$R^1\bar{f}_*\Omega^\bullet_{\bar{E}/\bar{X}}(d\log D/C) \to R^2\bar{f}_*(\Omega^{\bullet-1}_{\bar{E}/\bar{X}}(d\log D/C) \otimes \bar{f}^*\Omega^1_{\bar{X}}(d\log C))$$

that is,

$$\nabla : \mathcal{H} \to \mathcal{H} \otimes \Omega^1_{\bar{X}}(d\log C)$$

The locally free sheaf $\mathcal{H}_n = \mathrm{Sym}^n\mathcal{H}$ of rank $n+1$ inherits a "tensor product" Hodge filtration $(\mathrm{Fil}^i_n)_i$ and a "tensor product" log-connection ∇_n; the filtration (Fil^i_n) $(= i = 0, \ldots, n+1)$ is given by $\mathrm{Fil}^i_n\mathcal{H}_n = \sum_{k_1,\ldots,k_n} \mathrm{Fil}^{k_1}\mathcal{H} \otimes \cdots \otimes \mathrm{Fil}^{k_n}\mathcal{H}$ the sum being taken over all n-uples of integers k_j's, equal to 0, 1 or 2, such that $k_1 + \cdots + k_n = i$. It follows easily from the definition that $gr^i\mathcal{H}_n = \omega^{-n+2i}$ for $i = 0, \ldots, n$, and $gr^i\mathcal{H}_n = 0$ otherwise. Similarly, the log-connection $\nabla_n : \mathcal{H}_n \to \mathcal{H}_n \otimes \Omega^1_{\bar{X}}(d\log C)$ is given by $\nabla_n(v_1 \otimes \cdots \otimes v_n) = \sum_{i=1}^n v_1 \otimes \cdots \otimes \nabla(v_i) \otimes \cdots \otimes v_n$. It obviously satisfies Griffiths transversality: $\nabla_n(\mathrm{Fil}^i_n) \subset \mathrm{Fil}^{i-1}_n$. We denote by $\mathcal{H}(n)$ the de Rham complex $\nabla_n : \mathcal{H}_n \to \mathcal{H}_n \otimes \Omega^1_{\bar{X}}(d\log C)$ concentrated in degrees n and $n+1$. It is filtered by the absolute Hodge filtration $\mathrm{Fil}^i_{\mathrm{abs}}$ $(i = 0, \ldots, n+2)$ defined as the convolution of the "filtration bête" on $\mathcal{O}_{\bar{X}} \to \Omega_{\bar{X}}(d\log C)$ with $\mathrm{Fil}^\bullet_n$, namely:

$$\mathrm{Fil}^i_{\mathrm{abs}}\mathcal{H}(n) : \mathrm{Fil}^i_n \to \mathrm{Fil}^{i-1}_n \otimes \Omega_{\bar{X}}(d\log C)$$

for $i = 0, \ldots, n+1$ and $\mathrm{Fil}^{n+2}_{\mathrm{abs}}\mathcal{H}(n) = 0$. Hence, it gives rise to a sequence of complexes $\mathrm{gr}^i_{\mathrm{abs}}\mathcal{H}(n)$ (with differential $\mathrm{gr}^i_{\mathrm{abs}}\nabla n$):

$$\begin{array}{llcl}
\mathrm{gr}^0_{\mathrm{abs}}: & \omega^{-n} & \to & 0 \\
\mathrm{gr}^1_{\mathrm{abs}}: & \omega^{-n+2} & \to & \omega^{-n}\otimes\Omega_{\bar{X}}(d\log C) \\
\vdots & & & \\
\mathrm{gr}^i_{\mathrm{abs}}: & \omega^{-n+2i} & \to & \omega^{-n+2i-2}\otimes\Omega_{\bar{X}}(d\log C) \\
\vdots & & & \\
\mathrm{gr}^n_{\mathrm{abs}}: & \omega^{n} & \to & \omega^{n-2}\otimes\Omega_{\bar{X}}(d\log C) \\
\mathrm{gr}^{n+1}_{\mathrm{abs}}: & 0 & \to & \omega^{n}\otimes\Omega_{\bar{X}}(d\log C)
\end{array}$$

We then remark that for $i = 1, \ldots, n$, the differential $gr^i\nabla_n$ of the ith complex $\mathrm{gr}^i_{\mathrm{abs}}\mathcal{H}(n)$ is an isomorphism. More precisely, the construction of the Kodaira-Spencer homomorphism $KS : \omega \to \omega^{-1}\otimes\Omega^1_{\bar{X}}(d\log C)$ from the Gauss-Manin connection shows that $\mathrm{gr}^i_{\mathrm{abs}}\nabla_n = \mathrm{Id}_{\omega^{-n+2i-1}}\otimes KS$. Since KS is an isomorphism, the same holds for $\mathrm{gr}^i\nabla_n$.

Let us make these data explicit.

Recall the situation for $n = 0$. For each $\mathbb{Z}[\frac{1}{N}]$-affine neighborhood V of the divisor C of cusps in $\bar{X}$, we consider the affine covering (X, V) of $\bar{X}$; we form the Cech bicomplex:

$$\begin{array}{ccc}
\mathcal{O}_{\bar{X}}(X\cap V) & \overset{d}{\to} & \Omega^1_{\bar{X}}(X\cap V) \\
\uparrow & & \uparrow \\
\mathcal{O}_{\bar{X}}(X)\oplus\mathcal{O}\bar{X}(V) & \overset{d}{\to} & \Omega^1_{\bar{X}}(X)\oplus\Omega^1_{\bar{X}}(d\log C)(V)
\end{array}$$

The associated total complex (in degrees 0, 1, 2) calculates $H^1_{\log-dR}(\bar{X})$.

Similarly, for $n > 0$, the total complex (in degrees $n, n+1, n+2$) associated to the Cech bicomplex

$$\begin{array}{ccc}
\mathcal{H}_n(X\cap V) & \overset{\nabla_n}{\to} & \mathcal{H}_n\otimes\Omega^1_{\bar{X}}(X\cap V) \\
\uparrow & & \uparrow \\
\mathcal{H}_n(X)\oplus\mathcal{H}_n(V) & \overset{\nabla_n}{\to} & \mathcal{H}_n\otimes\Omega^1_{\bar{X}}(X)\oplus\mathcal{H}_n\otimes\Omega^1_{\bar{X}}(d\log C)(V)
\end{array}$$

calculates the $\mathbb{Z}[\frac{1}{N}]$-module $H^\bullet_{\log-dR}(\bar{X},\mathcal{H}_n)$. This cohomology module is filtered by the total Hodge filtration on the complex $\mathcal{H}(n)$ defined by the subcomplexes (concentrated in degrees n and $n+1$) $\mathrm{Fil}^i\mathcal{H}(n) = \mathrm{Fil}^i\mathcal{H}_n\oplus\mathrm{Fil}^{i-1}\mathcal{H}_n\otimes\Omega^1_{\bar{X}}(d\log C)$ for $i = 0, \ldots, n+1$; moreover, for $i = 1, \ldots, n$, $gr^i\mathcal{H}(n) = \omega^{-n+2i}\oplus\omega^{-n+2i-2}\otimes\Omega^1_{\bar{X}}(d\log C)$, while $gr^0\mathcal{H}(n) = \omega^{-n}\oplus 0$ and $gr^{n+1}\mathcal{H}(n) = 0\oplus\omega^n\otimes\Omega^1_{\bar{X}}(d\log C)$ We can now determine the spectral sequence associated to this filtration

$$E_1^{i,j} = H^{i+j}(\bar{X}, gr^i\mathcal{H}(n)) \Rightarrow H^{i+j-n}_{\log-dR}(\bar{X},\mathcal{H}_n)$$

The determination above of $gr^i\nabla_n$ leads to the

Corollary 2.1. *We have a natural short exact sequence of* $\mathbb{Z}[\frac{1}{N}]$*-modules*

$$0\to E_1^{n+1,0}\to H^1_{\log-dR}(\bar{X},\mathcal{H}_n)\to E_1^{0,n+1}\to 0$$

where $E_1^{n+1,0} = H^0(\bar{X}, \omega^{n+2})$ *and* $E_1^{0,n+1} = H^1(\bar{X}, \omega^{-n})$. *Here natural means compatible to algebraic correspondences.*

Proof. Indeed, since $gr^i\nabla_n$ are isomorphisms for $i = 0, \dots, n$, we notice that $E_1^{i,j} = 0$ unless $i = 0$ and $j = n+1$ or $i = n+1$ and $j = 0$, in which case $E_1^{n+1,0} = H^1(\bar{X}, gr^{n+1}\mathcal{H}(n)) = H^0(\bar{X}, \omega^n \otimes \Omega^1_{\bar{X}}(d\log C)) = H^0(\bar{X}, \omega^{n+2})$ by Kodaira-Spencer isomorphism. Similarly, $E_1^{0,n+1} = H^1(\bar{X}, gr^0\mathcal{H}(n))$ which is equal to $H^1(\bar{X}, \omega^{-n})$.

3. de Rham cohomology and dual BGG complex

Let p be a prime number not dividing $2N$ and such that $n+1 < p-1$. We fix as base ring the $\mathbb{Z}[\frac{1}{N}]$-algebra $\mathbb{Z}_p$. In this section all modules and sheaves of modules are $\mathbb{Z}_p$-modules. We recall how to compute the Hodge filtration on $\mathcal{H}_n$ and on $H^1_{\log-dR}(\bar{X}, \mathcal{H}_n)$ in group-theoretic terms, via the dual BGG complex.

3.1 *Relative Koszul and BGG complexes*

For any affine $\mathbb{Z}_p$-group scheme $H = \operatorname{Spec} A$, let $\mathfrak{h}$ be its Lie algebra, $U\mathfrak{h}$ its universal enveloping algebra, and $\mathcal{U}(H)$ its distribution algebra ([21] Chapter 7, or [12] II.1.12). Let J be the kernel of the counit character $A \to \mathbb{Z}_p$; recall that $\mathcal{U}(H) = \cup_k \mathcal{U}_k(H)$ where $\mathcal{U}_k(H) = \operatorname{Hom}_{\mathbb{Z}_p}(A/J^{k+1}, \mathbb{Z}_p)$. The product (of convolution) of two distributions $\mu, \nu : A \to \mathbb{Z}_p$ is the composition of their tensor product $\mu \otimes \nu : A \otimes A \to \mathbb{Z}_p$ with the comultiplication $c : A \to A \otimes A$. Since $\mathcal{U}_1(H) = \operatorname{Hom}_{\mathbb{Z}_p}(A/J^2, \mathbb{Z}_p) = \mathbb{Z}_p \oplus \mathfrak{h}$, there is a canonical filtered algebras homomorphism $U(\mathfrak{h}) \to \mathcal{U}(H)$; it is an isomorphism over a field of characteristic zero, hence induces an injective $\mathbb{Z}_p$-algebra homomorphism $U\mathfrak{h} \hookrightarrow \mathcal{U}(H)$. Its reduction modulo p, however, is not injective: over a field of characteristic p, the kernel is the ideal $U\mathfrak{h}^{[p]}$ generated by pth powers (see [21] II.1.12). This implies therefore that for any $k < p$, the injection $U_k\mathfrak{h} \hookrightarrow \mathcal{U}_k(H)$ is actually an isomorphism.

For any representation V of H defined over $\mathbb{Z}_p$, the natural action of $U\mathfrak{h}$ on V extends naturally to $\mathcal{U}(H)$. Namely, consider the comorphism $c_V : A(V) \to A \otimes_{\mathbb{Z}_p} A(V)$ of the action $H \times V \to V$ and for any $\mu \in \mathcal{U}(H)$, let us form $(\mu \otimes Id) \circ c_V$; this is the comorphism of the action of μ on V.

Recall also the Invariance Theorem ([12] II.4.6) that the ring of left-invariant differential operators on H is isomorphic to $\mathcal{U}(H)$.

If H is a split Chevalley group over $\mathbb{Z}_p$, one can give a rather explicit description of the inclusions $U\mathfrak{h} \subset \mathcal{U}(H) \subset U\mathfrak{h} \otimes \mathbb{Q}_p$. Namely, let $B = (Z_i, H_j, X_k)$ be a Chevalley basis of $\mathfrak{h}$ (with Z_i central, H_j semisimple and X_k nilpotent) then, $\mathcal{U}(H)$ is generated by the divided powers $\frac{X_k^n}{n!}$, and the divided difference products $\binom{Z_i}{n}$ and $\binom{H_j}{n}$ for all $n \geq 0$ (see [21], II.1.12 and [9] or [3] Chapt. VIII, Sect. 12).

Consider the $\mathbb{Z}_p$-group scheme $G = GL_2$; let T be its standard torus, consisting of diagonal matrices, and B its standard Borel, consisting of upper triangular matrices in G. Let V be the standard two dimensional representation of G and $V_n = \mathrm{Sym}^n V$; under the assumption $n < p$, this module can be called unambiguously the irreducible $\mathbb{Z}_p$-representation of highest weight n (see [29] Sect. 1.9, Lemma. Let $\mathfrak{g}$, $\mathfrak{b}$ resp. $\mathfrak{t}$ be the $\mathbb{Z}_p$-Lie algebra of G, B resp. of T. For any $\mathbb{Z}_p$-representations Y of G and Y' of B, recall the Garland-Lepowsky "tensor identity" (see [15] Prop. 1.7 and [29] Sect. 2.2). It provides a canonical isomorphism of $U\mathfrak{g}$-modules (and similarly of $\mathcal{U}(G)$-modules):

$$(TI1)\quad (U\mathfrak{g} \otimes_{U\mathfrak{b}} Y') \otimes_{\mathbb{Z}_p} Y \cong U\mathfrak{g} \otimes_{U_\mathfrak{b}} (Y' \otimes_{\mathbb{Z}_p} Y|_\mathfrak{b})$$

resp.

$$(TI2)\quad (\mathcal{U}(G) \otimes_{\mathcal{U}(B)} Y') \otimes_{\mathbb{Z}_p} Y \cong \mathcal{U}(G) \otimes_{\mathcal{U}(B)} (Y' \otimes_{\mathbb{Z}_p} Y|_\mathfrak{b})$$

where $Y|_\mathfrak{b}$ denotes the $\mathfrak{b}$-module restriction of the action of $\mathfrak{g}$ on Y to $\mathfrak{b}$.

Consider the natural morphism $\mathfrak{g}/\mathfrak{b} \to U_\mathfrak{g} \otimes_{U\mathfrak{b}} \mathbb{Z}_p$ which sends $\bar{X}$ to $X \otimes 1$, where $X \in \mathfrak{g}$ is any lifting of $\bar{X} \in \mathfrak{g}/\mathfrak{b}$. It gives rise to a resolution by (relative) Verma modules of the trivial representation $\mathbb{Z}_p$, called the relative Koszul resolution (see [1] or [29] Sect. 2.2).

$$0 \to U_\mathfrak{g} \otimes_{U\mathfrak{b}} \mathfrak{g}/\mathfrak{b} \stackrel{d_0}{\to} U_\mathfrak{g} \otimes_{U\mathfrak{b}} \mathbb{Z}_p \stackrel{d_{-1}}{\to} \mathbb{Z}_p \to 0$$

where d_{-1} is induced by the augmentation character $U\mathfrak{g} \to \mathbb{Z}_p$.

By tensoring by a representation Y and applying the isomorphisms (TI1), one obtains a resolution of the $\mathfrak{g}$-representation Y (relative Koszul complex):

$$U\mathfrak{g} \otimes_{U\mathfrak{b}} (\mathfrak{g}/\mathfrak{b} \otimes_{\mathbb{Z}_p} Y|_\mathfrak{b}) \stackrel{d_0}{\to} U\mathfrak{g} \otimes_{U\mathfrak{b}} Y|_\mathfrak{b}$$

with the maps $d_0(\xi \otimes \bar{X} \otimes v) = \xi X \otimes v - \xi \otimes Xv$ and $d_{-1}(\xi \otimes v) = \xi v$.

For purpose of coherence with the Hodge filtration later, we consider this complex as placed in degree $-(n+1)$ and $-n$.

Replacing $U\mathfrak{h}$ by $\mathcal{U}(H)$ for $H = G, B$, and using (TI2) instead of (TI1), we get another relative Koszul complex. We shift it by n, so that it is now concentrated in degrees $-1-n$ and $-n$. The resulting complex is denoted $S(G, B, Y)$:

$$S(G, B, Y)\quad \mathcal{U}(G) \otimes_{\mathcal{U}(B)} (\mathfrak{g}/\mathfrak{b} \otimes Y|_\mathfrak{b}) \to \mathcal{U}(G) \otimes_{\mathcal{U}(B)} Y$$

of generalized relative Verma $\mathbb{Z}_p$-modules (in the terminology of [29] Sect. 4.1). Since we deal with $G = GL_2$, so that G/B is one-dimensional, it turns out that this complex is actually a resolution of the G-representation Y (by generalized relative Verma modules), although it is not so for higher dimensional Chevalley groups.

Let (X_-, X_+, H, Z) be the standard basis of the Lie algebra $\mathfrak{g}$ of G, with $X_- = \left(\begin{smallmatrix} 0 & 0 \\ 1 & 0 \end{smallmatrix}\right)$, $X_+ = \left(\begin{smallmatrix} 0 & 1 \\ 0 & 0 \end{smallmatrix}\right)$, $H = \left(\begin{smallmatrix} 1 & 0 \\ 0 & -1 \end{smallmatrix}\right)$ and $Z = \left(\begin{smallmatrix} 1 & 0 \\ 0 & 1 \end{smallmatrix}\right)$; let $S = \left(\begin{smallmatrix} 0 & 1 \\ 1 & 0 \end{smallmatrix}\right)$ and $W = \{1, S\}$; note that $ad(S)(H) = -H$; finally, let $\mathcal{Z}$ be the center of $\mathcal{U}(G)$; as mentioned previously, it is generated by $\binom{Z}{n} (n \geq 0)$ and the center $\mathcal{Z}'$ of $\mathcal{U}(G')$ (distribution algebra of the derived group G' of G);

Consider the (arithmetic) Harish-Chandra isomorphism

$$\gamma_\circ : \mathcal{Z}' \otimes \mathbb{Q}_p \cong \mathbb{Q}_p[H]^W = \mathbb{Q}_p\left[\frac{1}{2}H^2 + H\right],$$

$$z = P(H) + y{\cdot}X_+ \mapsto \gamma_\circ(z) = P(H),$$

where the action of the symmetry $S \in W$ is by $H \mapsto -2 - H$. The image by $\gamma_\circ$ of the Casimir element $C = X_+X_- + X_-X_+ + \frac{1}{2}H^2$ is $\frac{1}{2}H^2 + H$.

Recall that the infinitesimal character of a Verma module $\mathcal{M}$ over $\mathbb{Z}_p$ or of a $\mathfrak{g}$-irreducible $\mathbb{Z}_p$-representation Y is the character of $\mathcal{Z}$ giving its action on $\mathcal{M}$ resp. Y.

By the very definition of the Harish-Chandra isomorphism, the infinitesimal character $\chi_{(n;c)}$ of $\mathcal{M}_{(n;c)} = \mathcal{U}(G) \otimes_{\mathcal{U}(B)} \mathbb{Z}_p w_{n,c}$ for $w_{n,c}$ a vector such that $Hw_{n,c} = nw_{n,c}$, $Zw_{n,c} = cw_{n,c}$ and $X_+w_{n,c} = 0$, $(n, c \in \mathbb{Z})$, is given by $\chi_{(n;c)}(C) = \frac{1}{2}n^2 + n$, $\chi_{(n;c)}(Z) = c$.

Moreover, the Verma modules $\mathcal{M}_{(n;c)}$ and $\mathcal{M}_{(n';c)}$ have the same infinitesimal character if and only if $c' = c$ and $n' = n$ or $n' = -2 - n$; in particular, the infinitesimal character $\chi_Y : \mathcal{Z} \to \mathbb{Z}_p$ of an irreducible representation (Y, ρ_Y) of G is uniquely determined by the pair of eigenvalues $(\rho_Y(H); \rho_Y(Z))$ of (H, Z) on a highest weight vector $w \in Y$: if $\rho_Y(H) = n$ and $\rho_Y(Z) = c$, then $\chi_Y(C) = \frac{1}{2}n^2+n$ (and $\chi_Y(Z) = c$).

Let us describe explicitely the subcomplex of $S(G, B, V_n^\vee)$ on which $\mathcal{Z}$ acts by the infinitesimal character of $V_n^\vee$ (up to semisimplification). Via the description above, the infinitesimal character of V_n is given by the pair $(n; n)$ hence that of $V_n^\vee$ is described by $(n; -n)$. On the other hand, for any pair of integers $(m, c) \in \mathbb{Z}^2$ such that $c \equiv m \pmod 2$, let $W_{(m;c)}$ be the one-dimensional representation of T given by the character $\mathrm{diag}(t_1, t_2) \mapsto t_1^m (t_1t_2)^{\frac{c-m}{2}}$. Recall that $\mathrm{Hom}_{\mathcal{U}(G)}(\mathcal{U}(G) \otimes_{\mathcal{U}(B)} W_{(m;c)}, \mathcal{U}(G) \otimes_{\mathcal{U}(B)} W_{(m';c')})$ does not vanish if and only if $c = c'$ and $m = m'$ or $m = -m' - 2$, in which cases it is free of rank one (see for instance [16]). For $m' = n$ and $m = -n - 2$, a basis of this module is provided by $\xi \otimes 1 \mapsto \xi X_-^{n+1} \otimes 1$. Indeed, for any choice of a basis w_k of $W_{(k;c)}$, the map $w_{-n-2} \mapsto X_-^{n+1} \otimes w_n$ is B-linear To see this, one applies the formulae in $U\mathfrak{g} : HX_-^k = X_-^k H - 2kX_-^k$ and $X_+X_-^k = X_-^k X_+ + kX_-^{k-1}H - k(k-1)X_-^{k-1}$ in order to get $H \cdot (X_-^{n+1} \otimes w_n) = -(n+2)X_-^{n+1} \otimes w_n$ and $X_+ \cdot (X_-^{n+1} \otimes w_n) = 0$.

The complex $S(G, B, V_n^\vee)$ is filtered by complexes of Verma modules; let $S(G, B, V_n^\vee))_{\chi(n;-n)}$ be the subcomplex cut by the infinitesimal character $\chi_{(n;-n)}$ of $V_n^\vee$. It is given by the last line of the diagram below. Note that it consists in $\mathcal{U}(G)$-simple generalized Verma modules. Since $n < p$, we get a direct factor subcomplex (see [29] Sect.2.7, Corollary).

$$\begin{array}{ccc}
((\mathcal{U}(G) \otimes_{\mathcal{U}(B)} \mathfrak{g}/\mathfrak{b}) \otimes_{\mathbb{Z}_p} Y)_{\chi(n;-n)} & \to & ((\mathcal{U}(G) \otimes_{\mathcal{U}(B)} \mathbb{Z}_p) \otimes_{\mathbb{Z}_p} Y)_{\chi(n;-n)} \\
\| & & \| \\
((\mathcal{U}(G) \otimes_{\mathcal{U}(B)} W_{(-2;0)}) \otimes_{\mathbb{Z}_p} Y)_{\chi(n;-n)} & & ((\mathcal{U}(G) \otimes_{\mathcal{U}(B)} \mathbb{Z}_p) \otimes_{\mathbb{Z}_p} W_{(n;-n)})_{\chi(n;-n)} \\
\| & & \| \\
\mathcal{U}(G) \otimes_{\mathcal{U}(B)} W_{(-n-2;-n)} & \to & \mathcal{U}(G) \otimes_{\mathcal{U}(B)} W_{(n;-n)}
\end{array}$$

where the equality of $\mathcal{U}(G)$-modules

$$((\mathcal{U}(G)\otimes_{\mathcal{U}(B)} W_{(-2;0)})\otimes_{\mathbb{Z}_p} V_n^{\vee})_{\chi(n;-n)} = \mathcal{U}(G)\otimes_{\mathcal{U}(B)} W_{(-n-2;-n)}$$

follows from the formula, valid for any $\mathfrak{b}$-representation T:

$$(\mathcal{U}(G)\otimes_{\mathcal{U}(B)} T)_{\chi(n;-n)} = \mathcal{U}(G)\otimes_{\mathcal{U}(B)} T_{\chi(n;-n)},$$

together with the fact that only the last term of the filtration by $\mathfrak{b}$-submodules of $W_{(-2;0)}\otimes_{\mathbb{Z}_p} V_n^{\vee}$ corresponds to an action of $\mathcal{Z}$ by the character $\chi_{(n;-n)}$.

Similar remark for the equality

$$(\mathcal{U}(G)\otimes_{\mathcal{U}(B)} V_n^{\vee})_{\chi(n;-n)} = \mathcal{U}(G)\otimes_{\mathcal{U}(B)} W_{(n;-n)}$$

Moreover, the bottom horizontal arrow is an injection given by $\xi\otimes w \mapsto \xi X_-^{n+1}\otimes w$.

Definition 3.1. *The BGG complex $BGG(V_n^{\vee})$ is the filtered subcomplex of $S(G,B,V_n^{\vee})$ given by:*

$$BGG\ (V_n^{\vee}) : \mathcal{U}(G)\otimes_{\mathcal{U}(B)} W_{(-n-2;-n)} \to \mathcal{U}(G)\otimes_{\mathcal{U}(B)} W_{(n;-n)}$$

placed in degrees $-n$ and $-(n+1)$.

Comments. Its differential, which is of order $n+1$ is given by $\xi\otimes w_{-n-2}\mapsto \xi X_-^{n+1}\otimes w_n$. Note that as a $\mathfrak{b}$-representation, $V_n^{\vee}$ admits $W_{(n;-n)}$ as a subrepresentation by sending w_n to v_n so we get

$$i_0 : U\mathfrak{g}\otimes_{U\mathfrak{b}} W_{(n;-n)} \hookrightarrow U\mathfrak{g}\otimes_{U\mathfrak{b}} V_n^{\vee}, \quad 1\otimes w_n \hookrightarrow 1\otimes v_n.$$

However, $\mathfrak{g}/\mathfrak{b}\otimes V_n^{\vee}$ does not admit $W_{(-n-2;-n)}$ as $\mathfrak{b}$-subrepresentation. This apparent paradox is solved as follows. In fact, $U\mathfrak{g}\otimes_{U\mathfrak{b}}\mathfrak{g}/\mathfrak{b}\otimes V_n^{\vee}$ admits several copies of $W_{(-n-2;-n)}$ as $\mathfrak{t}$-subrepresentations, namely the lines generated by the vectors $X_-^i\otimes\bar{X}_-\otimes X_-^{n-i}v_n$ (for $i=0,\dots,n$) but a simple calculation shows that it admits only one copy of $W_{(-n-2;-n)}$ as $\mathfrak{b}$-submodule, namely the line generated by

$$\sum_{i=0}^{n} X_-^i\otimes\bar{X}_-\otimes X_-^{n-i}v_n$$

The meaning of this statement is that this vector is the unique linear combination of the $X_-^i\otimes\bar{X}_-\otimes X_-^{n-i}v_n$'s annihilated by X_+. In particular, we have a morphism of $U\mathfrak{g}$-modules

$$i_1 : U\mathfrak{g}\otimes_{U\mathfrak{b}} W_{(-n-2;-n)} \hookrightarrow U\mathfrak{g}\otimes_{U\mathfrak{b}}\mathfrak{g}/\mathfrak{b}\otimes V_n^{\vee}$$

given by $1\otimes w_{-n-2}\mapsto -\sum_{i=1}^n X_-^i\otimes\bar{X}_-\otimes X_-^{n-i}v_n$.

The inclusion of the BGG complex as a subcomplex of the relative Koszul complex is given by (i_1,i_0). Indeed one checks easily that

$$i_0(X_-^{n+1}\otimes w_n) = d_0(i_1(1\otimes w_{-n-2})).$$

We define the increasing H-filtration Fil_i^H $(i = 0, 1, \ldots, n+2)$ on $S(G, B, V_n^\vee)$ by putting Fil_i^H to be the submodule in

$$\mathcal{U}(G) \otimes_{\mathcal{U}(B)} (\mathfrak{g}/\mathfrak{b} \otimes_{\mathbb{Z}_p} V_n^\vee) \oplus \mathcal{U}(G) \otimes_{\mathcal{U}(B)} V_n^\vee$$

induced from $\mathcal{U}(B)$ to $\mathcal{U}(G)$ of the B-submodules sum of the H-eigenspaces corresponding to eigenvalues $\geq n + 2 - 2i$:

$$\mathrm{Fil}_i^H : \mathcal{U}(G) \otimes_{\mathcal{U}(B)} (\mathfrak{g}/\mathfrak{b} \otimes_{\mathbb{Z}_p} V_n^\vee)_{\geq n+2-2i} \to \mathcal{U}(G) \otimes_{\mathcal{U}(B)} (V_n^\vee)_{\geq n+2-2i}$$

For $i \neq 1, n+2$, its i-th graded complex gr_i^H is given by an isomorphism

$$\mathrm{gr}_i^H : \mathcal{U}(G) \otimes_{\mathcal{U}(B)} W_{(n-2i;-n)} \cong \mathcal{U}(G) \otimes_{\mathcal{U}(B)} W_{(n-2i;-n)}$$

while

$$\mathrm{gr}_1^H : 0 \to \mathcal{U}(G) \otimes_{\mathcal{U}(B)} W_{(n;-n)}$$

and

$$\mathrm{gr}_{n+2}^H : \mathcal{U}(G) \otimes_{\mathcal{U}(B)} W_{(-n-2i;-n)} \to 0$$

Moreover, this filtration induces the following very simple filtration on BGG $(V_n^\vee)$:

$$\mathrm{Fil}_0^H\,\mathrm{BGG} = 0, \quad \mathrm{Fil}_1^H\,\mathrm{BGG} : 0 \to \mathcal{U}(G) \otimes_{\mathcal{U}(B)} W_{(n;-n)} \quad \text{and}$$
$$\mathrm{Fil}_{n+2}^H\,\mathrm{BGG} = \mathrm{BGG}(V_n^\vee).$$

For this filtration, we have the same graded pieces as above

$$\mathrm{gr}_1^H\,\mathrm{BGG} : 0 \to \mathcal{U}(G) \otimes_{\mathcal{U}(B)} W_{(n;-n)}$$

and

$$\mathrm{gr}_{n+2}^H\,\mathrm{BGG} : \mathcal{U}(G) \otimes_{\mathcal{U}(B)} W_{(-n-2;-n)} \to 0$$

Note that the differential is zero for each of these complexes.

We shall relate this filtration to the Hodge filtration in the following section.

3.2 *(log-)de Rham and dual BGG complexes*

On $\mathbb{Z}_p^2$ (column vectors), there is a standard symplectic pairing given by the determinant, preserved by the special linear group G', and a standard filtration, preserved by B; similarly, on the sheaf $\mathcal{H}$ defined in Sect. 2.2, there is the Poincaré symplectic pairing and the Hodge filtration. Let us consider the $\mathbb{Z}_p$-scheme $\mathcal{T}_{\mathcal{H}} = \mathrm{Isom}_{\bar{X},\mathrm{sympl},\mathrm{fil}}(\mathcal{O}_{\bar{X}}^2, \mathcal{H})$, where the index "sympl" and "fil" refers to isomorphisms compatible both to the symplectic pairings and the filtrations; note that on the target module, we view these data via the identification $\mathcal{O}_{\bar{X}}^2 = \mathbb{Z}_p^2 \otimes \mathcal{O}_{\bar{X}}$ (see [27] Sect. 5.2 for more details).

Let us put $B' = B \cap G'$; the morphism $\mathcal{T}_{\mathcal{H}} \to \bar{X}$ defines a right B'-torsor. It is related to the (right) $\mathbb{G}_m$-torsor $\mathcal{T}_\omega$ that we defined in Sect. 2.1. Namely, there is a forgetful morphism $\mathcal{T}_{\mathcal{H}} \to \mathcal{T}_\omega$ obtained by restricting a filtered isomorphism to the Fil^1 of the standard filtration of $\mathcal{O}^2_{\bar{X}}$ (taking values in $\omega \subset \mathcal{H}$).

The torsor $\mathcal{T}_{\mathcal{H}}$ allows us to define a covariant sheafification functors $F_{\mathbb{Z}_p}$ from the category of finite free $\mathbb{Z}_p$-representations of B' to the category $\mathcal{FF}$ of finite locally free $\mathcal{O}_{\bar{X}}$-modules with filtration.

Note that we should consider representations of B, that is, we should include the action of the center, but we shall omit it for simplicity, so that we write W_m instead of $W_{(m;c)}$, by forgetting about the central character.

For any object W of $\mathrm{Rep}_{\mathbb{Z}_p}(B')$, we define $F_{\mathbb{Z}_p}(W)$ as the sheaf of sections of the bundle

$$\mathcal{T}_{\mathcal{H}} \overset{B'}{\times} W \to \bar{X}$$

Actually, in the sequel, we'll also consider the contravariant version of $F_{\mathbb{Z}_p}$:

$$F^\vee_{\mathbb{Z}_p} = \mathrm{Hom}_{\mathcal{O}_{\bar{X}}}(-, \mathcal{O}_{\bar{X}}) \circ F_{\mathbb{Z}_p}$$

which is exact on the full subcategory of B'-representations which are finite free over $\mathbb{Z}_p$. For such a representation W, we thus have

$$F^\vee_{\mathbb{Z}_p}(W) = \mathrm{Hom}_{\mathcal{O}_{\bar{X}}}(F_{\mathbb{Z}_p}(W), \mathcal{O}_{\bar{X}}).$$

Remark. Denoting by T the Levi quotient of B, and by $\mathbb{G}_m$ the "semisimple" subtorus of T, so that $B'/N = \mathbb{G}_m$, we see that if W is a $\mathbb{G}_m$-module, then the forgetful morphism above induces a vector bundle isomorphism

$$\text{(CompTors)} \quad \mathcal{T}_{\mathcal{H}} \overset{B'}{\times} W \cong \mathcal{T}_\omega \overset{\mathbb{G}_m}{\times} W.$$

We see immediately that $F_{\mathbb{Z}_p}(\mathbb{Z}_p) = \mathcal{O}_{\bar{X}}$ and more generally for any $n \in \mathbb{Z}$, $F_{\mathbb{Z}_p}(W_n) = \omega^n$, so that $F^\vee_{\mathbb{Z}_p}(W_n) = \omega^{-n}$. This can be viewed from the definition since it is obvious that $F_{\mathbb{Z}_p}(W_1) = \omega$; it can also be checked over $\mathbb{C}$: the pull-back of $\mathcal{T}_\omega$ to the upper half plane P is given by $P \times \mathbb{C}^\times$ with left action of the discrete group Γ given by the automorphic factor $j(\gamma, z)$, hence W_n is sent to ω^n.

By the Kodaira-Spencer isomorphism, this implies that $F_{\mathbb{Z}_p}(\mathfrak{g}/\mathfrak{b}) = T_{\bar{X}}(-\log C)$ namely, the subsheaf of the tangent bundle consisting of sections which, near the cusps, are multiple of $q\frac{d}{dq}$; it is also the dual sheaf of $\Omega_{\bar{X}}(d\log C)$.

For $n < p$, the $\mathbb{Z}_p$-representation V_n of G is defined unambiguously and we have by definition $F^\vee_{\mathbb{Z}_p}(V_n^\vee) = \mathcal{H}_n$.

We want to extend the functor $F_{\mathbb{Z}_p}$ to the category of induced modules $M(V) = U\mathfrak{g} \otimes_{U\mathfrak{b}} V$, resp. $\mathcal{M}(V) = \mathcal{U}(G) \otimes_{\mathcal{U}(B)} V$ (V a $\mathfrak{b}$-module of weights $0 \leq \ell < p$), by the same formula

$$F_{\mathbb{Z}_p}(M(V)) = \mathcal{T}_{\mathcal{H}} \overset{B}{\times} U\mathfrak{g} \otimes_{U\mathfrak{b}} V$$

resp.

$$F_{\mathbb{Z}_p}(\mathcal{M}(V)) = \mathcal{T}_{\mathcal{H}} \overset{B}{\times} \mathcal{U}(G) \otimes_{\mathcal{U}(B)} V$$

it will take values in the category of finite modules over the ring $\bar{\mathcal{D}}_{\bar{X}}$ of log C-differential operators, resp. $\tilde{\mathcal{D}}_{\bar{X}}$ of log C-divided power differential operators, on $\bar{X}$ (see [13] p. 217 and [27] Sect. 4.2).

For this, we need to recall some algebra.

Let $G = \mathrm{Spec} A$, hence $A = \mathcal{O}_G$; consider the diagram of short exact sequences

$$\begin{array}{ccccccccc} 0 & \to & I & \to & A \otimes_{\mathbb{Z}_p} A & \stackrel{m}{\to} & A & \to & 0 \\ & & \downarrow & & \downarrow & & \downarrow & & \\ 0 & \to & J & \to & A & \stackrel{\epsilon}{\to} & \mathbb{Z}_p & \to & 0 \end{array}$$

where m denotes the multiplication, ϵ the counit morphism and the vertical maps are $1 \otimes f - f \otimes 1 - \mapsto f - \epsilon(f)$, resp. $\epsilon \otimes Id$, resp. ϵ.

Recall that $\mathcal{U}(G)$ is the union of the submodules $\mathcal{U}_k(G) = \mathrm{Hom}_{\mathbb{Z}_p}(A/J^{k+1}, \mathbb{Z}_p)$.

The left A-module $\mathcal{D}$ of differential operators on G is the union of $\mathcal{D}_k = \mathrm{Hom}_A((A \otimes A)/I^{k+1}, A)$, where the left-$A$-module structure on $A \otimes A$ is by multiplication on the first factor. We have therefore a natural map $\Phi : \mathcal{U}(G) \to \mathcal{D}$ given by $\Phi(\xi)(f \otimes g) = f\xi(g)$ The image of Φ is precisely the subring of left-invariant differential operators: $\eta \in \mathcal{D}$ such that for the left translation by any $g \in G$, $L_g^* \circ \eta = \eta \circ (L_g^* \otimes Id)$.

Similarly, let us compute $\mathcal{U}(G) \otimes_{\mathcal{U}(B)} \mathbb{Z}_p$ as a left B-module.

Let $\mathcal{J}_B$ be the kernel of the map $\mathcal{O}_B \to \mathbb{Z}_p$ of evaluation at 1. The big cell $N^- \times B$ of G is the open set of matrices $\left(\begin{smallmatrix} a_{11} & a_{12} \\ a_{21} & a_{22} \end{smallmatrix}\right)$ such that $a_{11} \neq 0$. Hence, the restriction to this cell induces an inclusion with $\mathbb{Z}_p$-flat cokernel $\mathcal{O}_G \hookrightarrow \mathcal{O}_G\left[\frac{1}{a_{11}}\right] = \mathcal{O}_{N^-} \otimes \mathcal{O}_B$. Therefore $\mathcal{O}_G \cap (1 \otimes \mathcal{J}_B)$ is $\mathbb{Z}_p$-direct factor in $\mathcal{O}_G$ and we have a commutative diagram

$$\begin{array}{ccccc} 0 \to \mathrm{Hom}((\mathcal{O}_{N_-} \otimes \mathcal{J}_B)_{<0}, \mathbb{Z}_p) & \to & \mathrm{Hom}(\mathcal{O}_{N_-} \otimes \mathcal{J}_B, \mathbb{Z}_p) & \to & \mathrm{Hom}(\mathcal{O}_G \cap (\mathcal{O}_{N_-} \otimes \mathcal{J}_B), \mathbb{Z}_p) \\ \uparrow & & \uparrow & & \uparrow \\ 0 \to \mathrm{Hom}((\mathcal{O}_{N_-} \otimes \mathcal{O}_B)_{<0}, \mathbb{Z}_p) & \to & \mathrm{Hom}(\mathcal{O}_{N_-} \otimes \mathcal{O}_B, \mathbb{Z}_p) & \to & \mathrm{Hom}(\mathcal{O}_G, \mathbb{Z}_p) \\ \uparrow & & \uparrow & & \uparrow \\ 0 & \to & \mathrm{Hom}(\mathcal{O}_{N_-} \otimes 1, \mathbb{Z}_p) & = & \mathrm{Hom}(\mathcal{O}_{N_-}, \mathbb{Z}_p) \end{array}$$

where the upper line of vertical arrows are injections, and the lower are surjections. We wrote $(\mathcal{O}_{N^-} \otimes \mathcal{O}_B)_{<0}$ for the $\mathbb{Z}_p$-submodule of $\mathcal{O}_{N^-} \otimes \mathcal{O}_B$ where the discrete valuation $\mathrm{ord}_{a_{11}}$ at a_{11} is strictly negative. The exactness of the horizontal lines, resp. vertical lines come from the $\mathbb{Z}_p$-linear decomposition $\mathcal{O}_{N^-} \otimes \mathcal{O}_B = \mathcal{O}_G \oplus (\mathcal{O}_{N^-} \otimes \mathcal{O}_B) < 0$ resp. $\mathcal{O}_{N^-} \otimes \mathcal{O}_B = (\mathcal{O}_{N^-} \otimes \mathcal{J}_B) \oplus (\mathcal{O}_{N^-} \otimes 1)$ from which it follows that $(\mathcal{O}_{N^-} \otimes \mathcal{J}_B)_{<0} = (\mathcal{O}_{N^-} \otimes \mathcal{O}_B)_{<0}$.

Thus, $\mathcal{U}(G) \otimes_{\mathcal{U}(B)} \mathbb{Z}p = \cup_k \mathrm{Hom}(\mathcal{O}_{N^-}/\mathcal{J}_{N^-}^{k+1}, \mathbb{Z}_p)$. Note that we have the following $\mathbb{Z}_p$-linear decomposition:

$$\mathrm{Hom}\ (\mathcal{O}_{N^-}/\mathcal{J}_{N^-}^{k+1}, \mathbb{Z}_p) = \bigoplus_{\ell \leq k} \mathbb{Z}_p \frac{1}{\ell!} \bar{X}_-^{\ell}$$

where X_- denotes the linear form sending $f \in \mathcal{O}_{N^-}$ to the value at $0 \in \mathbb{G}_a$ of its derivative, once one uses the 1-parameter group $U_{-a} : \mathbb{G}_a \to N^-$ sending x to $\left(\begin{smallmatrix} 1 & 0 \\ x & 1 \end{smallmatrix}\right)$ to identify $\mathbb{G}_a$ to N^-. One can then describe explicitely the left action of B by the formulas $H\bar{X}_-^k = -2k\bar{X}_-^k$, $X_+\bar{X}_-^k = -k(k-1)\bar{X}_-^k$.

By what precedes, the product $\mathcal{T}_\mathcal{H} \times \mathrm{Hom}(\mathcal{O}_{N^-}/J_{N^-}^{k+1}, \mathbb{Z}_p)$ identifies to the bundle of left B-invariant log C-PD differential operators of order $\leq k$ over $\mathcal{T}_\mathcal{H}$. Hence, the contracted product $\mathcal{T}_\mathcal{H} \overset{B'}{\times} \mathrm{Hom}(\mathcal{O}_{N^-}/J_{N^-}^{k+1}, \mathbb{Z}_p)$ identifies to the bundle of log C-PD differential operators of order $\leq k$ on $\bar{X}$.

Once this is fixed, one can determine the image of $\mathcal{M}(\mathbb{Z}_p)$ and $\mathcal{M}(\mathfrak{g}/\mathfrak{q})$. As left $\mathcal{O}_{\bar{X}}$-Module, we have

$$F_{\mathbb{Z}_p}(\mathcal{M}(\mathbb{Z}_p)) = \bigoplus_k \frac{1}{k!} T_{\bar{X}}(-d\log C)^{\otimes k} = \tilde{\mathcal{D}}_{\bar{X}}$$

and similarly

$$F_{\mathbb{Z}_p}(\mathcal{M}(\mathfrak{g}/\mathfrak{b})) = \tilde{\mathcal{D}}_{\bar{X}} \otimes_{\mathcal{O}_{\bar{X}}} T_{\bar{X}}(-d\log C).$$

More generally, for any $\mathbb{Z}_p$-finite free B-module W (with fixed central action), put $F_{\mathbb{Z}_p}(W) = \mathcal{W}$; then

$$F_{\mathbb{Z}_p}(\mathcal{M}(W)) = \tilde{D}_{\bar{X}} \otimes_{\mathcal{O}\bar{X}} \mathcal{W}.$$

Remark. The rings $\mathcal{D}_{\bar{X}}$, $\bar{\mathcal{D}}_{\bar{X}}$ and $\tilde{\mathcal{D}}_{\bar{X}}$ are $\mathcal{O}_{\bar{X}}$-bimodules, but not $\mathcal{O}_{\bar{X}}$-algebras (non-commutativity of the functions with the differential operators).

Lemma 3.2.

(1) $F_{\mathbb{Z}_p}(M(\mathbb{Z}_p)) = \bar{\mathcal{D}}_{\bar{X}}$, $F_{\mathbb{Z}_p}(\mathcal{M}(\mathbb{Z}_p)) = \tilde{\mathcal{D}}_{\bar{X}}$ *and*

$$F_{\mathbb{Z}_p}(M(\mathfrak{g}/\mathfrak{b})) = \bar{\mathcal{D}}_{\bar{X}} \otimes_{\mathcal{O}_{\bar{X}}} T_{\bar{X}}(-d\log C),$$
$$F_{\mathbb{Z}_p}(\mathcal{M}(\mathfrak{g}/\mathfrak{b})) = \tilde{\mathcal{D}}_{\bar{X}} \otimes_{\mathcal{O}_{\bar{X}}} T_{\bar{X}}(-d\log C),$$

(2) *The image by $F_{\mathbb{Z}_p}^\vee$ of the relative Koszul resolution of $\mathbb{Z}_p : M(\mathfrak{g}/\mathfrak{b}) \to M(\mathbb{Z}_p)$ yields the log-de Rham complex $\mathcal{O}_{\bar{X}} \overset{d}{\to} \Omega_{\bar{X}}(d\log C)$.*

(3) *For any $n \in [0, p[$, the image by $F_{\mathbb{Z}_p}^\vee = \mathrm{Hom}_{\mathcal{O}_{\bar{X}}}(-, \mathcal{O}_{\bar{X}})$ of the relative Koszul resolution of $V_n^\vee : M(\mathfrak{g}/\mathfrak{b} \otimes V_n^\vee) \to M(V_n^\vee)$ yields the log-de Rham complex $\mathcal{H}(n)$ for $\mathcal{H}_n$ with its Gauss-Manin log-connection ∇_n:*

$$\mathcal{H}_n \overset{\nabla_n}{\to} \mathcal{H}_n \otimes_{\mathcal{O}_{\bar{X}}} \Omega_{\bar{X}}(d\log C)$$

Proof. 1) is already proved. For 2) and 3), let us briefly recall how from a morphism of degree 1

$$(*) \quad \bar{\mathcal{D}}_{\bar{X}} \otimes_{\mathcal{O}_{\bar{X}}} \mathcal{W}_1 \to \bar{\mathcal{D}}_{\bar{X}} \otimes_{\bar{X}} \mathcal{W}_0$$

gives rise to a (log-)connection $\mathcal{W}_0^\vee \to \mathcal{W}_1^\vee \otimes_{\bar{X}} \Omega_{\bar{X}}(d\log C)$.

Let $\bar{\mathcal{D}}_{\bar{X},i}$, $i = 0$ resp. 1, be the module of log C-differential operators of degree 0, resp. at most 1. Since the morphism $(*)$ has degree 1, it preserves $\bar{\mathcal{D}}_{\bar{X},1}$. We have $\mathcal{O}_{\bar{X}} = \bar{\mathcal{D}}_{\bar{X},0}$ while $\bar{\mathcal{D}}_{\bar{X},1}$ inserts in a short exact sequence of $\mathcal{O}_{\bar{X}}$-modules

$$0 \to T_{\bar{X}}(-d\log C) \stackrel{a}{\to} \bar{\mathcal{D}}_{\bar{X},1} \stackrel{b}{\to} \mathcal{O}_{\bar{X}} \to 0$$

By composing with the inclusion $a \otimes Id_{\mathcal{W}}$ and the projection $b \otimes Id_{\mathcal{W}}$, the map $(*)$ gives rise to

$$T_{\bar{X}}(-d\log C) \otimes \mathcal{W}_1 \to \mathcal{W}_0$$

whose dual provides the desired log-connection.

Let us apply this to the image by $F_{\mathbb{Z}_p}$ of $M(\mathfrak{g}/\mathfrak{b} \otimes V_n^\vee) \to M(V_n^\vee)$. The morphism is given, for $\xi \in U\mathfrak{g}$, $\bar{Y} \in \mathfrak{g}/\mathfrak{b}$ and $v^\vee \in V_n^\vee$, by $\xi \otimes \bar{Y} \otimes v^\vee \mapsto \xi Y \otimes v^\vee - \xi \otimes Y v^\vee$ hence it is degree 1, and after dualizing we find a morphism

$$\delta : \mathcal{H}_n \to \mathcal{H}_n \otimes_{\mathcal{O}_{\bar{X}}} \Omega_{\bar{X}}(d\log C)$$

given for $f \in \mathcal{O}_{\bar{X}}$ and $v \in \mathcal{H}_n$ by $fv \mapsto (\bar{Y} \otimes v^\vee \mapsto Yf \cdot \langle v, v^\vee\rangle - f\langle Yv, v^\vee\rangle)$, in this formula we recognize the Gauss-Manin connection ∇_n as desired. QED.

We can also define a similar functor $F_{\mathbb{Z}/p\mathbb{Z}}$ for the categories of $\mathbb{Z}/p\mathbb{Z}$-representations resp. $\mathbb{Z}/p\mathbb{Z}$-generalized Verma modules. By construction, we have

$$F_{\mathbb{Z}/p\mathbb{Z}}(W/pW) = F_{\mathbb{Z}_p}(W) \otimes \mathbb{Z}/p\mathbb{Z}, \quad \text{and} \quad F_{\mathbb{Z}/p\mathbb{Z}}(\mathcal{M}/p\mathcal{M}) = F_{\mathbb{Z}_p}(\mathcal{M}) \otimes \mathbb{Z}/p\mathbb{Z}.$$

It sends $S(G, B, V_n^\vee)$ to the logarithmic de Rham complex on $\bar{X}$.

Let $\mathcal{K}^\bullet(n) = (\mathcal{K}_n^0 \to \mathcal{K}_n^1)$ be the subcomplex of $\mathcal{H}(n)$ defined as the image of the complex BGG $(G, B, V_n^\vee)$ by $F_{\mathbb{Z}_p}^\vee$ (resp. $F_{\mathbb{Z}/p\mathbb{Z}}$). The complex $\mathcal{K}^\bullet(n)$ is called the dual BGG complex.

One has $\mathcal{K}_n^0 = \omega^{-n}$ and $\mathcal{K}_n^1 = \omega^{n+2}$ with a differential induced by ∇_n. By Definition 3.1, this differential is dual to $w_{-n-2} \mapsto X_-^{n+1} \otimes_{w_n}$. If we want to take into account the determinant character, we put $\omega^s(t) = \mathcal{W}_{s;s+2t}$; then, the exact formula for the dual BGG complex is

$$(**) \quad \mathcal{K}_n^\bullet : \omega^{-n}(n+1) \to \omega^{n+2}$$

This is important for the future action of the Hecke operators on the complex.

Theorem 2. *The complex $K^\bullet(n)$ is a direct factor filtered subcomplex of $H^\bullet(n)$ which is quasi-isomorphic to $H^\bullet(n)$.*

The Hodge filtration on $H^\bullet(n)$ induces on $K^\bullet(n)$ the filtration

$$\begin{array}{lccc} \mathrm{Fil}^0 : & \omega^{-n} & \to & \omega^{n+2} \\ \mathrm{Fil}^{n+1} : & 0 & \to & \omega^{n+2} \\ \mathrm{Fil}^{n+2} : & 0 & \to & 0 \end{array}$$

The proof is given for more general groups in [27] Theorem 6 Sect. 5.4. This is where one needs to introduce the divided power versions of the Koszul resp. BGG *complexes (that is to replace $U\mathfrak{g}$ by $U(G)$), and also to introduce a functor, called the L functor, sending $\bar{D}_{\bar{X}}$ to $\tilde{D}_{\bar{X}}$, and the logarithmic de Rham complex to its divided power analogue in order to transform the log de Rham complex and the BGG dual complex into résolutions.*

3.3 *The ordinary locus*

We use the notations of Section 2.1. Let $\bar{X}\otimes\mathbb{Z}/p\mathbb{Z}$ be the modulo p modular curve and $\mathcal{G} \to \bar{X}\otimes\mathbb{Z}/p\mathbb{Z}$ the reduction modulo p of the semi-abelian curve, neutral component of the universal generalized elliptic curve. Let $\mathcal{G}[p]^0$ be the neutral component of the quasi-finite group scheme $\mathcal{G}[p]$ of p-torsion in $\mathcal{G}$. It is a finite flat group scheme over $\bar{X} \otimes \mathbb{Z}/p\mathbb{Z}$ The ordinary locus S of $\bar{X} \otimes \mathbb{Z}/p\mathbb{Z}$ is the largest subscheme over which $\mathcal{G}[p]^0$ is of multiplicative type.

The multiplication by p on $\mathcal{G}$ factors through the relative Frobenius morphism $F : \mathcal{G} \to \mathcal{G}^{(p)}$, where $\mathcal{G}^{(p)}$ denotes the pull-back of $\mathcal{G}$ by the absolute Frobenius on $\bar{X} \otimes \mathbb{Z}/p\mathbb{Z}$. Let V (the Verschiebung morphism) be the quotient morphism $V : \mathcal{G}^{(p)} \to \mathcal{G}$. Recall that for any line bundle $\mathcal{L}$ on $\bar{X} \otimes \mathbb{Z}/p\mathbb{Z}$, there is a canonical isomorphism $\mathcal{L}^{(p)} \cong \mathcal{L}^{\otimes p}$ The Hasse invariant H is the modulo p modular form of weight $p-1$ defined as the morphism $V^* : \omega \to \omega^{(p)}$ deduced from V. By definition of S, this is the locus where V is étale, that is, where H does not vanish. Hence S is open. It is non-empty hence dense, by existence of the Tate curve. Since ω is ample on $\bar{X} \otimes \mathbb{Z}/p\mathbb{Z}$, the open subscheme S is affine.

3.4 *Frobenius and* U_p

Let $\bar{X}_0(p)$ be the (compactified) modular curve classifying p-isogenies $(E \to E')$ over $\bar{X}$ over $\mathbb{Z}_p$ (see [24]). Note that by definition of $\bar{X}$, an auxiliary $\Gamma_1(N)$-level structure ($N \geq 4$) with p prime to N, is present but not mentioned in the sequel. We have two degeneracy maps π_1 sending $(E \to E')$ to E and π_2 sending $(E \to E')$ to E'. An algebraic correspondence (π_1, π_2, α) acting on the pair $(\bar{X}, \mathcal{F})$ for a sheaf $\mathcal{F}$ on $\bar{X}$ is determined by (π_1, π_2) and a morphism $\alpha : \pi_2^*\mathcal{F} \to \pi_1^*\mathcal{F}$ For $\mathcal{F} = \mathcal{H}_n$, the correspondence T_p acts on $H^\bullet_{dR}(\bar{X}, \mathcal{H}_n)$ by $\pi_{1,*} \circ \alpha_* \circ \pi_2^*$ where $\alpha : \pi_2^*\mathcal{H}_n \to \pi_1^*\mathcal{H}_n$ is defined by $\mathrm{Sym}^n\left(\begin{smallmatrix} p & 0 \\ 0 & 1 \end{smallmatrix}\right) : V_n^\vee \to V_n^\vee$. Let us consider the "Frobenius" subcorrespondence ϕ obtained by restricting this diagram to the ordinary locus S of $\bar{X} \otimes \mathbb{Z}/p\mathbb{Z}$ and by restricting π_1 and π_2 to the multiplicative type part $\bar{X}^m$ of $\pi_1^{-1}(S)$ (one knows that the ordinary locus of $\bar{X}_0(p)$ is the disjoint union of two components: $\bar{X}^m$ where the kernel of the isogeny is of multiplicative type, and $\bar{X}^e$, where it is étale). This subcorrespondence is of degree p and is actually a purely inseparable morphism because π_2 is purely inseparable of degree p while π_1 is an isomorphism $\bar{X}^m \cong S$.

Let $\widehat{\bar{X}_{(p)}}$ be the formal scheme completion along the special fiber of $\bar{X}_{\mathbb{Z}_p}$ and S_∞ be its open formal subscheme with underlying scheme S. Let H the canonical subgroup $H \subset \mathcal{G}$ lifting the kernel of Frobenius in the universal (semi-)elliptic curve $\mathcal{G} \to S_\infty$ (see Sect. 2.1 for the notations); the Frobenius morphism ϕ lifts as a morphism $\phi : S_\infty \to S_\infty : \phi(x) = y$ if and only if $E_y = E_x/H_x$; this morphism is finite and flat.

It extends to the Igusa tower $T_{\infty,\infty} \to S_\infty$ as a finite flat endomorphism, so we can form ϕ_* as an endomorphism of the p-adic algebra $V(N)$ of p-adic modular forms; it leaves stable its submodules $H^0(S_\infty, \omega^k)$ for any $k \in \mathbb{Z}$.

We define U_p as $\frac{1}{p}\phi_*$. The q-expansion principle together with the formula $\sum_{n\geq 1} a_n(f|U_p)q^n = \sum_{n\geq 1} a_{np}(f)q^n$ implies that U_p preserves integrality hence preserves $V(N)$ and its submodules $H^0(S_\infty, \omega^k)$ for any $k \in \mathbb{Z}$.

If f is a classical form of level $\Gamma_1(N_p)$, it can be viewed as a p-adic modular form of level prime to p as usual (see Hida [18]). Then, the definition above of U_p is compatible with the definition of this operator on classical forms.

Finally, let us recall that we defined in Sect. 2 an integral version of the Eichler-Shimura morphism for $k = n + 2 \geq 2$:

$$H^0(\bar{X}, \omega^k) \hookrightarrow H^1_{\log -dR}(\bar{X}, \mathcal{H}_n)$$

Here, we view both modules over $\mathbb{Z}_p$. By functoriality, this morphism is Hecke-equivariant for all T_n's (n prime to N) including T_p. Moreover in the restriction morphism

$$\mathrm{Res} : H^0(\bar{X}_{\mathbb{Z}_p}, \omega^k) \to H^0(S_\infty, \omega^k),$$

$\mathrm{Res}(f|T_p)$ is congruent to $\mathrm{Res}(f)|U_p$ mod. p because $k \geq 2$.

4. Comparison theorem and companion forms in *p*-small weight $k \geq 2$

Let f be our weight k new form of level N. Recall that we fixed a p-adic embedding $\bar{\mathbb{Q}} \hookrightarrow \bar{\mathbb{Q}}_p$; let $K \subset \bar{\mathbb{Q}}_p$ be a finite extension of $\mathbb{Q}_p$, with ring of integers $\mathcal{O}$ containing the eigenvalues of f such that the Galois representation ρ_f is defined over $\mathcal{O}$; let $\mathbb{F} = \mathcal{O}/(\varpi)$ be its residue field. Let V_f be the rank 2 free $\mathcal{O}$-module of this representation. Recall its cohomological definition. For $A = \mathbb{Q}_p, \mathbb{Z}_p, K, \mathcal{O}, \mathbb{F}$, let $V_n(A) = \mathrm{Symm}^n A^2$ viewed as a representation of G over A and $H^1(A) = H^1_{\mathrm{et}}(X \otimes \mathrm{Spec}\, \bar{\mathbb{Q}}, V_n(A))$; it is a $\mathrm{Gal}(\bar{\mathbb{Q}}/\mathbb{Q})$-representation defined over A. Let $V_f(K)$ be the 2-dimensional vector space cut by the idempotent 1_f associated to f in the Hecke algebra of level N and weight k (which acts faithfully on $H^1(K)$). Then, define V_f as $V_f(K) \cap H^1(\mathcal{O})$. By definition, it is a direct factor in $H^1(\mathcal{O})$. Under the assumptions that $(p, N) = 1$ and $n + 1 < p - 1$, we know that $H^i(\mathcal{O})$ is torsion-free for $i = 0, 1, 2$. This implies that $H^1(\mathcal{O})/\varpi H^1(\mathcal{O}) \cong H^1(\mathbb{F})$ and that $V_f \hookrightarrow H^1(\mathcal{O})$ induces an injection $\bar{V}_f = V_f/\varpi V_f \hookrightarrow H^1(\mathbb{F})$.

4.1 *The integral comparison theorem*

Under the assumption $n + 1 < p - 1$, Falting's log-crys comparison theorem states that, for $A = \mathbb{Q}_p, \mathbb{Z}_p, \mathbb{F}_p$, the contravariant Fontaine-Laffaille functor D sends the restriction to $D_p = \mathrm{Gal}(\bar{\mathbb{Q}}_p/\mathbb{Q}_p)$ of $H^1(A)$ to the log-crystalline cohomology module $H^1_{lc}(A) = H^1_{\log -\mathrm{crys}}(\bar{X}/A, \mathcal{H}(n))$, viewed as a ϕ-filtered A-module. Moreover, as a filtered A-module, $H^1_{\log -\mathrm{cr}}(\bar{X}/A, \mathcal{H}(n))$ coincides with $H^1_{\log -\mathrm{dR}}(\bar{X}/A, \mathcal{H}(n))$. Moreover, by this functor, the D_p-submodule V_f, resp. $\bar{V}_f$, of $H^1(\mathcal{O})$ resp. $H^1(\mathcal{O})$, is sent to the ϕ-filtered submodule M_f resp. $\bar{M}_f$ of $H^1_{lc}(\mathcal{O}) = H^1_{lc}(\mathbb{Z}_p) \otimes \mathcal{O}$ resp. $H^1_{lc}(\mathbb{F}) = H^1_{lc}(\mathbb{F}_p) \otimes \mathbb{F}$.

The Hodge filtration on $H^1_{lc}(A)$ has been determined in Sect. 2 using the quasi-isomorphism between the de Rham and the dual BGG complex. We found $H^1_{lc}(A) = H^1_{\log-\mathrm{cr}}(\bar{X}/A, \mathcal{H}(n)) = H^1_{\log-\mathrm{cr}}(\bar{X}/A, \mathcal{K}(n))$, so $\mathrm{Fil}^0 = \cdots = \mathrm{Fil}^n = H^1_{lc}(A)$, $\mathrm{Fil}^{n+1} = \mathrm{Im}(H^0(\bar{X}_A, \omega^{n+2}) \to H^1_{lc}(A))$ and $\mathrm{Fil}^{n+2} = 0$.

Since M_f is a sub-filtered module of $H^1_{lc}(\mathcal{O})$, we see that $\mathrm{Fil}^{n+1} M_f$ is $\mathcal{O}$-free of rank one, generated by the image $[f]$ of the differential form $\omega_f = (2i\pi)^{n+2} f(z) dv^{n+2}$ defined over $\mathcal{O}$ (as usual in this notation, the variable z is the uniformizing variable of the complex modular curve and v is the uniformizing of the elliptic curve above z). By Scholl's theorem [30], the ($\mathcal{O}$-linear) crystalline Frobenius ϕ on M_f is annihilated by $(X - \alpha)(X - \beta) \in \mathcal{O}[X]$. Moreover, it induces two maps $\phi_0 : \mathrm{Fil}^0(\bar{M}_f) \to \bar{M}_f$ and $\phi_{n+1} : \mathrm{Fil}^{n+1}(\bar{M}_f) \to \bar{M}_f$ the latter being induced by $p^{-(n+1)}\phi|\mathrm{Fil}^{n+1}$. Recall that the weak admissibility condition on $\bar{M}_f$ reads

$$\bar{M}_f = \phi_0(\mathrm{Fil}^0(\bar{M}_f)) + \phi_{n+1}(\mathrm{Fil}^{n+1}(\bar{M}_f))$$

Moreover, as explained in Perrin-Riou's paper [28], on the filtered module side, the ordinarity of the representation V_f at p is equivalent to the following compatibility between the Hodge filtration $\mathrm{Fil}^\bullet(M_f)$ and the slope filtration. Recall that the steps of the slope filtration $\mathrm{Fil}^\bullet_\phi$ are defined by $\mathrm{Fil}^i_\phi(M_f)$ is the sum of the ϕ-eigenspaces where the slope is $\geq i$. In this 2-dimensional situation, let (e_1, e_2) an $\mathcal{O}$-basis of M_f made of ϕ-eigenvectors: $\phi(e_1) = \alpha e_1$ and $\phi(e_2) = \beta e_2$; then, the compatibility reads simply as

$$M_f = \mathrm{Fil}^{n+1}(M_f) \oplus \mathcal{O} \cdot e_1$$

Moreover the decomposability assumption of $\bar{V}_f$ translates as the decomposability of $\bar{M}_f$, that is,

$$\mathrm{Fil}^{n+1}(\bar{M}_f) = \mathbb{F} \cdot \bar{e}_2$$

Under this assumption, we see that $[f] \in \mathrm{Fil}^{n+1}(\bar{M}_f)$ happens to be an eigen-vector for ϕ_{n+1}, with eigenvalue the reduction modulo ϖ of $\frac{\beta}{p^{n+1}}$. That is, $\phi_{n+1}([f]) - \frac{\beta}{p^{n+1}}[f] = 0$ in $H^1_{lc}(\mathbb{F})$. In order to exploit this formula we need to introduce the ordinary locus of $\bar{X}_{\mathbb{F}_p}$.

4.2 *Dual BGG and companion forms*

Recall that the invertible sheaf ω on $\bar{X}_{\mathbb{Z}_p}$ is relatively ample. Let H be the Hasse invariant on $\bar{X}_{\mathbb{F}_p}$; it is a non zero global section of ω^{p-1}; therefore, the locus S where it does not vanish is a non-empty affine open subset of $\bar{X}_{\mathbb{F}_p}$. It is called the ordinary locus because its geometric points correspond to (generalized) elliptic curves which are ordinary. Since it is affine, we have

- $H^i(S, \mathcal{F}) = 0$ for every $i > 0$ and for any coherent sheaf $\mathcal{F}$ on S, and
- $H^i(S, \mathcal{F}^\bullet) = H^i(\mathcal{F}^\bullet(V))$ for any complex $\mathcal{F}^\bullet$ of coherent sheaves.

Applying this to the dual BGG complex, we find that

$$H^1(S, \mathcal{H}(n)) = H^1(S, \mathcal{K}(n)) = \omega^{n+2}(S)/\nabla_n(\omega^{-n}(S)).$$

Moreover, the Frobenius morphism ϕ_S on S induces an automorphism of $H^0(S, \omega^m)$ for any m, in such a way that the short exact sequence

$$Ex(S)\ 0 \to H^0(S, \omega^{-n}) \to H^0(S, \omega^{n+2}) \to H^1(S, \mathcal{H}(n)) \to 0$$

is ϕ-equivariant. Let $H^0(]S[, \omega^m)$ be the rigid cohomology of the affinoid tube $]S[$.

Hence, the differential form $\eta = \frac{\phi}{p^{k-1}}(\omega_f) - \frac{\beta}{p^{k-1}}\omega_f \in H^0(S, \omega^k)$ maps to $[\eta] = 0$ in $H^1(S, \mathcal{H}(n))$. By exactness of $Ex(S)$, there exists $\omega' \in H^0(S, \omega^{-n}(n+1))$ such that $\eta = \nabla_n(\omega')$ (here, we mention the twist by $n+1$, because it becomes important in what follows).

A priori, ω' is only defined over S. Let us first assume it is eigen for U_p with a p-adic unit eigenvalue; this assumption is motivated by a formula of Coleman whose proof is recalled in [4] Sect. 3, $\phi = p^{k-1}\langle p\rangle U_p^*$ on $]S[$, so $\frac{\phi}{p^{k-1}}(\omega_f) - \frac{\beta}{p^{k-1}}\omega_f =$ (unit) $\cdot\, \omega_{f-af(pz)}$; the form $f - \alpha f(pz)$ is eigen for U_p with eigenvalue β. Since $p^{k-1}d \circ U_p = U_p \circ d$ (for the BGG dual differential d), we would find $\omega'|U_p = \frac{\beta}{p^{k-1}} \cdot \omega'$ if we could divide by p^{k-1} (that is, in characteristic zero), and ω' would be ordinary. Thus, by Jochnowitz's lemma (or Hida Theory [20]), the form $H \cdot \omega' \in H^0(S, \omega^{p-1-n})$, being of weight $k' = p + 1 - k \geq 2$ and ordinary, would extend to the whole curve $\bar{X}_{\mathbb{F}}$. Moreover, by Hida Theory [20], it would even be associated to the reduction modulo p, say $\bar{g}$, of a characteristic zero ordinary cusp eigenform g of weight $k' : \nabla_n(\omega_{\bar{g}}) = \omega'$.

However, this argument does not go through because we are in characteristic p so that the ordinarity does not seem to follow from the information above. However, let us prove that $\omega' H$ does extend to the whole curve $\bar{X}_{\mathbb{F}}$. Let $t \geq 1$ be the order of pole of the meromorphic form ω' along the supersingular divisor $Z \subset \bar{X}$. We need to see that $t \leq 1$. The endomorphism $\frac{\phi}{p^{k-1}}$ of $H^0(V, \omega^k)$ maps $H^0(\bar{X}, \omega^k)$ to $H^0(\bar{X}, \omega^{pk})$; therefore, by the q-expansion principle applied to $H^0(\bar{X}, \omega^{pk})$, we have the following equality of forms in $H^0(\bar{X}, \omega^{pk})$:

$$\bar{\omega}_f^{(p)} - \delta_1 H^k \bar{\omega}_f = H^k d\omega'$$

where $\bar{\omega}_f^{(p)} = \frac{\phi}{p^{k-1}}\bar{\omega}_f$. The left-hand side has no pole along Z. By a calculation of Katz [23], the operator d increases the order of pole at a supersingular point x by $k-1$: if $\mathrm{ord}_x(\omega') = t$, $\mathrm{ord}_x(d\omega') = t + k - 1$. By a theorem of Igusa, $\mathrm{ord}_x(H) = 1$, hence the right hand side has a pole of order at x equal to $t - 1$. We must therefore have $t \leq 1$. That is, $\omega' H$ extends to the whole curve modulo p.

On the other hand, ω' belongs to the localization $H^0(V, \omega^{-n}(n+1))_{\mathfrak{m}}$; since $H \equiv 1 \pmod{p}$, the multiplcation by H commutes to Hecke operators outside N_p, so that $\omega' H$ belongs to the localization $H^0(\bar{X}, \omega^{p-1-n}(n+1))_{\mathfrak{m}}$. This artinian module is non zero, hence its socle (that is, the part annihilated by $\mathfrak{m}$) is also non zero. In other words, we can assume that $H\omega'$ is eigen for the twisted action of the Hecke algebra.

Recall that for any prime ℓ different from p, we do have $\ell^{k-1} d \circ T_\ell = T_\ell \circ d$. Therefore, after untwisting, we see that the eigenvalues of $H\omega'$ are the characteristic p numbers $b_\ell = \frac{a_\ell}{\ell^{k-1}}$. By Deligne-Serre's lemma, this eigensystem lifts to characteristic zero so that there exists an eigenform g of weight $p+1-k$ and a prime $\mathfrak{p}$ above p in the field of coefficients of g, so that the eigenvalues B_ℓ of g satisfy

$$B_\ell \equiv \frac{a_\ell}{\ell^{k-1}} \pmod{\mathfrak{p}} \quad \text{for all } \ell \neq p.$$

We conclude by irreducibility of $\bar{\rho}_f$ that the Galois representation $\bar{\rho}_f \otimes \omega^{k-1}$ is modular:

$$\bar{\rho}_f \otimes \omega^{k-1} = \bar{\rho}_g$$

5. Overconvergent companion forms in weight 1

In their work on the degree 2 Artin-Langlands conjecture, Buzzard and Taylor for their construction of a weight one cusp eigenform f realizing a given degree two Artin representation, made the assumption that this representation is residually modular. More precisely, they proved the following

Theorem 3. *Let $p \geq 5$ and let $\mathcal{O}$ be a discrete valuation ring with uniformizing parameter ϖ which is finite and flat over $\mathbb{Z}_p$; let $\rho : \mathrm{Gal}(\bar{\mathbb{Q}}, \mathbb{Q}) \to \mathrm{GL}_2(\mathcal{O})$ be a continuous representation such that*

- *ρ is unramified outside a finite set of primes*
- *$\rho \pmod{\varpi}$ is modular and absolutely irreducible*
- *ρ is unramified at p and $\rho(Frob_p)$ has eigenvalues α and β which are distinct modulo ϖ,*

then, there exists a modular form of weight one and an embedding $i : \mathbb{Q}(f) \hookrightarrow \mathcal{O}$ such that $\rho_f = \rho$. In particular, ρ has finite image and $L(\rho, s) = L(f, s)$ is entire

Their method of proof consists in using these assumptions to construct two distinct Hida families; the authors then specialize these two families in weight one Then, they proceed to glue the overconvergent weight one eigenforms thus obtains, to construct the sought for classical eigenform f.

In the subsequent section, we explain how in the construction of the two families and their specialization in weight one as overconvergent forms, we may avoid the assumption of unramifiedness at p and the use of weight 2 and, sometimes, the use of a level divisible by p, by allowing a higher weight $k \in [2, p[$. We don't discuss the deepest problem, namely, the analytic continuation of these overconvergent forms.

5.1 *Galois deformations and modularity*

Let $p \geq 5$ be a prime; let f be a cusp eigenform of level N prime to p, character ϵ_f; assume that f is p-ordinary and of weight $k \in [2, p[$. Let $\mathcal{O}$ be the ring of integers of a p-adic field containing the eigenvalues of f such that $a_p(f) = \alpha + \beta$ with $\alpha \in \mathcal{O}^\times$ and $\alpha\beta = p^{k-1}\epsilon_f(p)$; let ϖ a uniformizing parameter of $\mathcal{O}$ and $\kappa = \mathcal{O}/(\varpi)$ its residue field.

Assume that the residual representation $\bar{\rho}_f$ splits at p:

$$\bar{\rho}_f|D_p = \xi(\bar{\alpha}) \oplus \xi\left(\overline{\frac{\beta}{p^{k-1}}}\right)\omega^{1-k}$$

Let us consider the deformation problem over local artinian $\mathcal{O}$-algebras with residue field κ, which send such an algebra A to the set of strict conjugacy classes of Galois representations $\rho : \mathrm{Gal}(\bar{\mathbb{Q}}/\mathbb{Q}) \to \mathrm{GL}_2(A)$ deforming $\bar{\rho}_f : \mathrm{Gal}(\bar{\mathbb{Q}}/\mathbb{Q}) \to \mathrm{GL}_2(\kappa)$ such that

- ρ is unramified outside N_p,
- ρ is ordinary at p; moreover, its restriction to D_p is of the form

$$\rho|D_p \sim \begin{pmatrix} \xi(\alpha_p) & * \\ 0 & * \end{pmatrix}$$

  with $\alpha_p \equiv \alpha \pmod{\mathfrak{m}_A}$

This deformation problem is prorepresentable over a universal deformation ring R_α^{univ}. By a theorem of Diamond (in weight 2) and R. Ramakrishna's thesis (for generalization to higher weights $k \in [2, p[$), R_α^{univ} is a "natural" quotient of Hida's universal ordinary Hecke algebra $h^0(N)$ with auxiliary level N; by "natural", we mean, as on p. 911 of [7], that the quotient map sends T_ℓ resp. S_ℓ to $\rho^{\mathrm{univ}}(\mathrm{Frob}_\ell)$ resp. $\det \rho^{\mathrm{univ}}(\mathrm{Frob}_\ell)$ for ℓ prime to N_p, resp. U_ℓ to 0 for ℓ dividing N, and sends U_p to $\alpha^{\mathrm{univ}} \in (R_\alpha^{\mathrm{univ}})^\times$, where all the Frobeniuses are geometric and where α^{univ} is the scalar by which Frob_p acts on the unramified line $H^0(I_p, \rho^{\mathrm{univ}})$. This theorem is actually a direct application of the Taylor-Wiles method (using Ramakrishna's thesis) if one assumes that the deformations are minimal at primes ℓ dividing N (see Taylor-Wiles, or see [32]). In particular, the representation ρ_f provides the classical eigenform of weight k and level N_p : $f'\alpha, k = f - \beta f(pz)$ which satisfies $f_{\alpha,k}|U_p = \alpha f_\alpha, k$.

But, more interestingly, given an Artin representation $\rho : \mathrm{Gal}(\bar{\mathbb{Q}}/\mathbb{Q}) \to \mathrm{GL}_2(\mathcal{O})$ such that

$$\rho|D_p \sim \begin{pmatrix} \xi(\alpha_p) & 0 \\ 0 & \xi(\beta_p)\omega^{1-k} \end{pmatrix}$$

and $\bar{\rho} = \bar{\rho}_f$, there exists a character $\lambda_\alpha : h^0(N) \to \mathcal{O}$ which factors through R^{univ}, say, as $\lambda_\alpha : R^{\mathrm{univ}} \to \mathcal{O}$ and such that $\rho_\alpha^{\mathrm{univ}} \otimes_{R_\alpha^{\mathrm{univ}}, \lambda_\alpha} \mathcal{O} = \rho$. This provides by Hida theory a p-adic cusp eigenform f_α of weight 1, such that $f_\alpha|U_p = \alpha_p \cdot f_\alpha$. This form being p-ordinary is overconvergent (as explained in [?] Lemma 1, Sect. 2), hence defines a section of the sheaf ω on an affinoid neighborhood of the rigid ordinary locus $]S[$ in $\bar{X}^{\mathrm{rig}}$.

On the other hand, by the theorem on companion forms proven above, there exists a p-ordinary cusp eigenform g of level N and weight $k' = p + 1 - k$ in $[2, p[$, such that $\bar{\rho} \otimes \omega^{1-k} = \bar{\rho}_g$.

Let us consider the deformation problem which sends a local $\mathcal{O}$-algebra A as before to the set of strict conjugacy classes of Galois representations $\rho : \mathrm{Gal}(\bar{\mathbb{Q}}/\mathbb{Q}) \to \mathrm{GL}_2(A)$ deforming $\bar{\rho}_g : \mathrm{Gal}(\bar{\mathbb{Q}}/\mathbb{Q}) \to \mathrm{GL}_2(\kappa)$ such that

- ρ is unramified outside N_p,
- ρ is ordinary at p and its restriction to D_p is of the form

$$\rho|D_p \sim \begin{pmatrix} \xi(\beta_p) & * \\ 0 & * \end{pmatrix}$$

with $\beta_p \equiv \frac{\beta}{p^{k-1}} \pmod{\mathfrak{m}_A}$

The representation $\rho \otimes \omega^{k-1}$ provides such a deformation of $\bar{\rho}_g$. The universal deformation ring R_β^{univ} is again a "natural" quotient of $h^0(N)$. This is defined by the same condition for T_ℓ, S_ℓ, U_ℓ for $\ell \neq p$, but at p the condition is that U_p maps to $\beta^{\mathrm{univ}} \in (R_\beta^{\mathrm{univ}})^\times$ such that $\beta^{\mathrm{univ}} \equiv \frac{\beta}{p^{k-1}} \pmod{\mathfrak{m}_\beta^{\mathrm{univ}}}$.

Again, this implies there exists a character of $\mathcal{O}$-algebras $\lambda_\beta : h^0(N) \to \mathcal{O}$ which factors through R_β^{univ}, say, as $\lambda_\beta : R_\beta^{\mathrm{univ}} \to \mathcal{O}$, such that $\rho_\beta^{\mathrm{univ}} \otimes_{R_\beta^{\mathrm{univ}}, \lambda_\beta} \mathcal{O} = \rho$. This character gives rise to a p-adic cusp eigenform f_β with eigenvalues $\lambda_\beta(T_\ell)$ for ℓ prime to N_p and such that $f_\beta|U_p = \beta_p f_\beta$.

Thus, we have proven:

Proposition 5.1. *Given a degree two odd Artin representation $\rho : \mathrm{Gal}(\bar{\mathbb{Q}}/\mathbb{Q}) \to \mathrm{GL}_2(\mathcal{O})$ finitely ramified, such that*

$$\rho|D_p \sim \begin{pmatrix} \xi(\alpha_p) & 0 \\ 0 & \xi(\beta_p)\omega^{1-k} \end{pmatrix}$$

and which is residually irreducible and such that $\bar{\rho} = \bar{\rho}_f$ for a cusp eigenform of level N prime to p and weight $k \in [2, p[$, there exist two p-adic overconvergent cusp eigenforms f_{α_p} and f_{β_p} of weight 1 and auxiliary level N such that $f_{\alpha_p}|U_p = \alpha_p \cdot f_{\alpha_p}$, resp. $f_{\beta_p}|U_p = \beta_p \cdot f_{\beta_p}$.

This result has been recently generalized to the $GS_p(4, \mathbb{Q})$-case in [32] and [33] when one replaces degree two odd Artin representations by degree four symplectic p-adic representations $\rho_{A,p}$ associated to irreducible abelian surfaces over $\mathbb{Q}$. The proof follows the same plan as the $\mathrm{GL}_2(\mathbb{Q})$-case: one constructs a companion form g to a p-ordinary cusp eigenform f of level N prime to p and p-small weight (k_1, k_2) (that is, $k_1 \geq k_2 \geq 3$ and $k_1 + k_2 - 3 < p - 1$), under the assumption that its residual representation $\bar{\rho}_f$ "partially splits" at p. Then, one considers two deformation problems consisting in p-ordinary symplectic rank four representations deforming $\bar{\rho}_f$ resp. $\bar{\rho}_g$. Then, we use a result of the type $R = T$ generalizing Diamond's theorem, but only in the minimal deformation case. Its specialization associated to $\rho_{A,p}$ resp. $\rho_{A,p}\otimes\omega^{k_2-2}$ yields the two overconvergent p-adic cusp eigenforms f_{α_A} resp. f_{β_A} of weight (2, 2) analogue to those in the proposition above.

However, the next point in the Artin case, treated in [7] and [8], is to make use of these forms to produce a classical cusp eigenform f_1 of level N or N_p such that $\rho_{f_1}=\rho$. In the $GS_p(4,\mathbb{Q})$-case, this question remains open.

References

[1] I. N. Bernstein, I. M. Gel'fand and S. I. Gel'fand, Differential operators on the base affine space, and a study of $\mathfrak{g}$-modules, in Lie Groups and Their Representations, ed. I. M. Gel'fand, Conf. Budapest 1971, Adam Hilger Publ. (1975).

[2] P. Berthelot, Cohomologie rigide et cohomologie à supports propres, Prépublication IRMAR (1996) 96–03.

[3] N. Bourbaki, Groupes et algèbres de Lie, Chapitres 7 et 8, Hermann (1975).

[4] C. Breuil and M. Emerton, Représentations p-adiques ordinaires de $GL_2(\mathbb{Q}_p)$ et compatibilité local-global, to appear in Astérisque.

[5] S. Bosch, Lectures on Formal and Rigid Geometry, Heft 378, preprint reihe SFB 478, Univ. Muenster (2005).

[6] S. Bosch, U. Güntzer and R. Remmert, Non-archimedean Analysis. A systematic approach, Grundlehren der Mathematischen Wissenschaften, Bd 261, Springer Verlag (1984).

[7] K. Buzzard and R. Taylor, Companion forms and weight one forms, *Ann. of Math.*, **149** (1999) 905–919.

[8] K. Buzzard, Analytic continuation of overconvergent eigenforms, *J. Amer. Math. Soc.*, **16** (2003) 29–55.

[9] C. Chevalley, Certains schémas de groupes semi-simples, Séminaire Bourbaki, Exposé, **219** (1960–1961).

[10] Robert F. Coleman, Classical and overconvergent modular forms, *Inv. Math.*, **124** (1996) 215–241.

[11] Robert F. Coleman and F. Voloch, Companion forms and Kodaira-Spencer theory, *Inv. Math.*, **110** (1992) 263–281.

[12] M. Demazure and P. Gabriel, Groupes algébriques, tome 1, Masson et Cie, Editeurs, North-Holland Publishing Co. (1970).

[13] G. Faltings and C.-L. Chai, Degeneration of abelian varieties, *Erg. der Math. u. ihre Grenzgebiete,* Springer Verlag (1990).

[14] G. Faltings and B. Jordan, Crystalline cohomology and GL(2, $\mathbb{Q}$), *Isr. J. Math.*, **90** (1995) 1–66.

[15] H. Garland and J. Lepowsky, Lie algebra homology and the MacDonald formulas, *Inv. Math.*, **34** (1976) 37–76.

[16] P. Garrett, Verma modules, Harish-Chandra homomorphisms, course notes, Sept., 14 (2006), http://www.math.umn.edu/garrett/

[17] B. Gross, A tameness criterion for Galois representations associated to modular forms (mod. p), *Duke Math. J.*, **61** (1990) 445–517.

[18] H. Hida, p-adic automorphic forms on reductive groups, in Automorphic Forms, *I, Astérisque*, **298** (2005).

[19] H. Hida, Control theorems of coherent sheaves on Shimura varieties of PEL type, *J. Inst. Math. Jussieu*, **1** (2002) no. 1, 1–76.

[20] H. Hida, Geometric Modular Forms and Elliptic Curves, World Scientific Publishing Co. (2000).

[21] J. C. Jantzen, Representations of algebraic groups, Academic Press 1987 or second edition, *Amer. Math. Soc.* (2003).

[22] B. W. Jordan and R. Livné, Integral Hodge Theory and congruences between modular forms, *Duke Math. J.*, **80** (1995) 419–484.

[23] N. Katz, p-adic properties of modular schemes and modular forms, in Modular Forms of One Variable III, pp. 69–189, Springer Lect. Notes 350, Springer (1972).

[24] N. Katz and B. Mazur, Arithmetic Moduli of Elliptic Curves, *Ann. of Math.*, **108**, Studies Princeton University Press (1985).

[25] M. Kisin and K. F. Lai, Overconvergent Hilbert Modular Forms, *Amer. J. Math.*, **127** (2005) no. 4, 735–783.

[26] A. Mokrane, Quelques remarques sur l'ordinarité, *J. Number Theory*, **73** (1998) 162–181.

[27] A. Mokrane and J. Tilouine, Cohomology of Siegel varieties with p-adic in tegral coefficients and applications, *Astérisque*, **280**, Publ. Soc. Math. France (2002) 1–95.

[28] B. Perrin-Riou, Représentations p-adiques ordinaires, Périodes p-adiques (Bures-sur-Yvette, 1988), *Astérisque*, **223** (1994) 185–220.

[29] A. Polo and J. Tilouine, Bernstein-Gelfand-Gelfand complexes and cohomology of nilpotent over $\mathbb{Z}_{(p)}$ for representations with p-small weights, *Astérisque*, **280**, Publ. Soc. Math. France (2002) 97–135.

[30] A. Scholl, Motives for modular forms, *Inv. Math.*, **100** (1990) 419–430.

[31] J.-P. Serre, Endomorphismes complètement continus des espaces de Banach p-adiques, *Publ. Math. de l'I.H.E.S.*, tome **12** (1962) 69–85.

[32] J. Tilouine, Nearly ordinary degree four symplectic Galois representations and p-adic Siegel modular forms, *Compos. Math.*, **142** (2006) 1122–1156.

[33] J. Tilouine, Companion forms and BGG complex, to appear in Ann. Inst. Fourier (2011)

[34] H. Yoshida, Siegel's Modular Forms and the Arithmetic of Quadratic Forms, *Inv. Math.*, **60** (1980) 193–248.

Made in United States
Orlando, FL
04 November 2026

90200855R00083